PRINCIPLES AND APPLICATIONS IN NUCLEAR ENGINEERING - RADIATION EFFECTS, THERMAL HYDRAULICS, RADIONUCLIDE MIGRATION IN THE ENVIRONMENT

Edited by **Rehab O. Abdel Rahman**
and **Hosam El-Din M. Saleh**

Principles and Applications in Nuclear Engineering - Radiation Effects, Thermal Hydraulics, Radionuclide Migration in the Environment
http://dx.doi.org/10.5772/65990
Edited by Rehab O. Abdel Rahman and Hosam El-Din M. Saleh

Contributors

Bingqi Zhu, David Ryan, Roger Saint-Fort, Roland Wisniewski, Jianjun Xu, Tianzhou Xie, Xuezhi Zhang, Lei Yang, Sannappa J., Rehab O. Abdel Rahman

First published in London, United Kingdom, 2018 by IntechOpen
IntechOpen is the global imprint of INTECHOPEN LIMITED, registered in England and Wales, registration number: 11086078, The Shard, 25th floor, 32 London Bridge Street
London, SE19SG – United Kingdom
Printed in Croatia

British Library Cataloguing-in-Publication Data
A catalogue record for this book is available from the British Library

Additional hard copies can be obtained from orders@intechopen.com

Principles and Applications in Nuclear Engineering - Radiation Effects, Thermal Hydraulics, Radionuclide Migration in the Environment, Edited by Rehab O. Abdel Rahman and Hosam El-Din M. Saleh
p. cm.
Print ISBN 978-1-78923-616-3
Online ISBN 978-1-78923-617-0

For EU product safety concerns:
IN TECH d.o.o., Prolaz Marije Krucifikse Kozulić 3, 51000 Rijeka, Croatia,
info@intechopen.com or visit our website at intechopen.com.

Meet the editors

Rehab O. Abdel Rahman is an Associate Professor of Chemical Nuclear Engineering at Atomic Energy Authority of Egypt. She received her PhD degree in Nuclear Engineering from Alexandria University. She has published more than 30 peer-reviewed scientific papers, 10 book chapters, and 5 books. She is an honored scientist of ASRT, serves as a verified reviewer for several journals, and is a managing editor of IJEWM and IJEE.

Hosam El-Din M. Saleh is a Professor of Radioactive Waste Management at the Radioisotope Department, Atomic Energy Authority, Egypt. He received his MSc and PhD degrees in Physical Chemistry from Cairo University. He is interested in studying innovative, economic, and environment-friendly techniques for the management of hazardous and radioactive wastes. He has authored many peer-reviewed scientific papers and chapters. He is a book editor of different books related to valuable international publishers.

Contents

Preface

The wide spread of the technological applications of nuclear sciences in our daily life is supported by the development in nuclear engineering. This engineering field of study was initiated to address engineering aspects of fission and fusion reactions and their applications in different sectors. It relies on the integration of knowledge gained from nuclear chemistry and physics, material engineering, mechanical, and electric power engineering. Specialized studies emerged from this integration include but not limited to nuclear safety, neutronics, radiation protection, radiological detection, and nuclear waste management. This book aims at presenting principles and applications in three important topics in nuclear engineering, namely; radiation effects, thermal hydraulics, and radionuclide migration in the environment. The authors summarized their experiences and present advances in different fields related to the presented topics. This book targets professionals in the nuclear industry and readers with technical backgrounds such as graduate and postgraduate students undertaking courses in nuclear chemistry and chemical and nuclear engineering.

The book consists of four sections that cover important research and development efforts in nuclear technologies. The first section introduces the nuclear engineering research areas and the role of nuclear safety in ensuring the sustainability of the nuclear industry with a special reference to this role in nuclear waste management. The editors introduced basic safety concepts and principals in nuclear waste management activities. This section presents the nuclear safety role in supporting the decision-making process and highlights this role for nuclear disposal project by listing requirements for safety cases and safety assessments.

The second section is concerned with radiation effects on materials, where principles and advanced investigations of radioactivity phenomenon were summarized. Prof. Sannappa demonstrated the current knowledge about radiation types and sources and the application of radioisotopes to support different industrial sectors. In addition, the chapter presents principals of radiation interaction with biological systems, radiation protection, and occupational and public control measures. An overview on advanced investigations of g interactions with different dense gas-metal systems was presented by Prof. Wiśniewski. This chapter summarized the results of more than a decade of cooperation between the Institute of Nuclear Research in Dubna and the National Center for Nuclear Research in Poland to study the interaction of g rays (10-23MeV) in Pd-D_2, Pd-Re-D_2, and pure gases, i.e.,H_2, D_2, and He.

The third section presents two aspects of nuclear thermal hydraulics, namely thermal performance of dense granular materials as high power targets and two-phase flow in narrow channels. Prof. Zhang and Prof. Yang addressed the advances in assessing the thermal hydraulic performance for dense granular materials for their potential use as targets in accelerator-driven systems and irradiation facilities. They presented an overview on heat transfer media with a special reference to the thermal properties of granular flow. The results of implementation of these media in prototype testing facilities were presented. Prof. Jianjun et al. demonstrated experimental and theoretical studies on boiling heat transfer in narrow rectangle channels. The experimental observations of bubble growth rate, bubble departure diameter, and bubble interface parameter are reported, and the analysis of bubble growth, departure mechanisms and slide and lift-off behaviors was studied by analyzing the force balance on growing bubbles.

The last section, Radionuclide Migration in Environmental Systems, presents principles and advanced applications in this field. Prof. Saint-Fort overviewed radioactivity principals and chemical characteristics of radionuclides and sorption theoretical concepts and models. Finally, the role of sorption in retarding radionuclide migration was introduced. Prof. Zhu and Prof. Ryan addressed an advanced application in uranium migration monitoring by studying the complexation of uranyl ions with dissolved organic matter using fluorescence excitation-emission matrix (EEM) techniques and regional integration analysis (RIA). The theoretical bases for RIA were introduced, and five sampling locations were selected in Merrimack River Valley in Massachusetts, USA, to conduct the study.

We would like to cordially thank all the authors for their efforts that led to the production of this distinguished scientific contribution. A special acknowledgment is directed to Ms. Maja Bozicevic, Author Service Manager, for her exceptional efforts.

Rehab O. Abdel Rahman and Hosam El-Din M. Saleh
Atomic Energy Authority of Egypt, Egypt

Section 1

Introduction

Chapter 1

Introductory Chapter: Safety Aspects in Nuclear Engineering

Rehab O. Abdel Rahman and Hosam M. Saleh

Additional information is available at the end of the chapter

http://dx.doi.org/10.5772/intechopen.76818

1. Introduction

In principle, engineering is the application of scientific knowledge to optimize production processes and product applications where the whole production life cycle is considered starting from design phase and ending by closure phase. Recent scientific knowledge of radiation phenomenon backs to more than a century, since then efforts were directed to understand and utilize this phenomenon to support human civilization [1]. Currently, radiation uses have been extended to support primary, secondary, and tertiary economical sectors; this includes but not limited to applications in oil and gas extraction, medical diagnosis procedures, power generation, and sterilization and gauging activities in health, industrial, and agricultural fields all over the globe [1, 2]. Ensuring optimum production and utilization of natural and induced radioactive materials to serve these sectors over extended periods of time was the reason to support the development of nuclear engineering sciences. In this respect, the output of the nuclear industry could be classified into four main products, namely, fission energy, radioisotopes, radiological detection instruments, and fusion energy. Subsequently, nuclear engineering could be viewed as the engineering field that is concerned with optimization of the processes that utilize and apply nuclear, fission and fusion reactions. Within all these processes, nuclear safety is addressed to guarantee safe and sustainable optimized productions and applications.

Research efforts that support the nuclear industry, as in any other industry, could be classified based on the maturity of the technology either commercial or innovative. These efforts are directed to improve the performance of the first class, whereas they aim at getting the second class into wide scale commercial applications [3]. Nuclear engineering research areas that cover the four identified industrial outputs could be listed as follows:

IntechOpen

1. Fission technologies which are concerned with design and optimization of nuclear research or power reactors, associated nonradiological systems and nuclear fuel cycle technologies.
2. Radioisotope production technologies, these include accelerators, irradiation facilities and their nonradiological systems, and radioisotope applications and associated waste management.
3. Radiological detection and nuclear instruments technologies, which include detection instruments that help in dose monitoring to ensure personal and facility safety.
4. Fusion technologies, which are innovative technologies that aim at engineering fusion energy.

Figure 1 shows different scientific bases that were used to establish nuclear engineering sciences and the areas that are covered within nuclear engineering studies. Specialized studies emerged from the integration of these scientific bases that include but not limited to neutronics, radiation protection, radiological detection, and nuclear waste management. This chapter focuses on introducing the role of nuclear safety in ensuring the sustainability of nuclear industry with special reference to nuclear waste management, where disposal activity is used as an example to illustrate this role. The reason for this selection is the nature of the disposal concept that relies on passive safety functions to ensure radiological hazard containment and confinement. Within this context, basic safety concepts and principles and nuclear waste management activities will be introduced. The role of nuclear safety in supporting decision-making process in the industry will be presented by highlighting this role for nuclear disposal project. Requirement for safety case and safety assessment for radioactive waste disposal

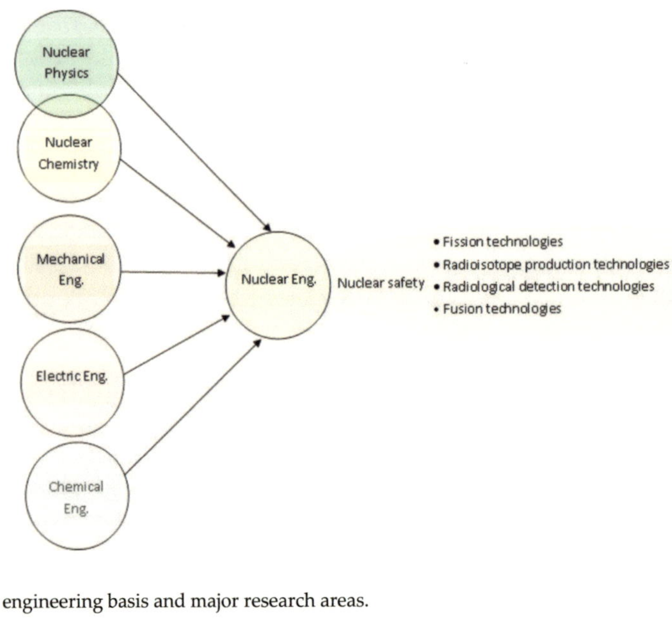

Figure 1. Nuclear engineering basis and major research areas.

project will be reviewed and the development of safety case throughout the disposal project life cycle will be introduced.

2. Safety concepts and principles

In principle, safety aims at ensuring protection from unwanted consequences. So, in workplaces that contain hazards, safety aims at controlling hazards to ensure acceptable risk. In facilities that handle radioactive materials, aside from the radiological hazard, that is, radiation and criticality, there could be several types of chemical and physical related hazards, that is, chemical toxicity, flammability, explicability and corrosivity [2]. These hazards and their classifications are listed as follows:

1. Radiation hazard: caused due to exposure to radioactive materials/radiation source, this exposure can lead to radiation health effect. The extent of these effects is dependent on the radiation type, absorbed dose, and duration of exposure and is initiated when living tissue absorbs some of the radiation energy that lead to changes in the cells [1, 4].
2. Criticality: in nuclear engineering, it refers to the capability of sustaining chain fission reaction; induced criticality should be assessed in workplaces that contain fissile materials [4].
3. Chemical toxicity: caused due to exposure to chemicals that can develop adverse reaction in living organisms [5]. Exposure could be classified based on the type, that is direct and indirect, or duration of exposure, that is acute, chronic [1, 5, 6].
4. Flammability, it is the ability of the material either solid or liquid or gas to ignite. Gas flammability is defined at standard pressure (101.3 kPa) and temperature (20°C). However, flammable liquids are classified based on their flash and boiling points (flashpoint <93°C). Solid flammable materials are those cause fire through friction [7].
5. Corrosivity, material that can damage metals or nonmetals, and it is classified based on the produced corrosion rate.
6. Explicability, materials that are chemically active producing gases at pressure rate that cause damage to its environment.

In this chapter, only nuclear safety will be discussed, where IAEA defined nuclear safety as "*the achievement of proper operating conditions, prevention of accidents or mitigation of accident consequences, resulting in protection of workers, the public and the environment from undue radiation hazards*" [4]. The acceptable levels of protection from radiological hazard is usually derived based on the recommendation of International Commission on Radiological Protection (ICRP) to keep the exposure probability and magnitude as low as reasonable achieved (ALARA) taking into account the economical and social factors [1, 4]. It should be noted that radiological protection should be ensured under both normal and incidental conditions, where low probability incidents with considerable radiological consequences should be considered. Safety measures should be taken to prevent and mitigate the consequence of incidents. Consequently, nuclear safety is only concerned with technical aspects to ensure protection from radiological hazard.

A decade ago, 3S (safety-security-safeguard) concept was introduced to ensure successful peaceful utilization of nuclear technology, where the areas of interaction between the three subconcepts, that is, safety, security, and safeguard, should be carefully addressed among concerned stakeholders [3, 8–10]. In this context, nuclear security is related to "prevention and detection of and, response to, theft, sabotage, unauthorized access, illegal transfer or other malicious acts involve nuclear material, other radioactive substances or their associated facilities" [4]. Safeguards are agreements between IAEA and member states that target fissile materials, where these materials should be declared and controlled.

International Atomic Energy Agency (IAEA) identified 10 principles to achieve the radiological protection objectives of nuclear safety under normal and incidental situations, as follows [9]:

1. Safety responsibility rests with operator either personal, or facility that might cause radiological risk.
2. Effective sustainable legislative framework should be established.
3. Effective sustainable leadership and management for safety should be established.
4. Benefits associated with any radiological practice should be balanced with associated risks.
5. Radiological protection must be optimized to provide the highest level of safety that could be reasonably achieved.
6. Radiation control measures must limit radiological risks to individuals.
7. Protection of environment and people from anticipated radiological risks should be ensured.
8. All safety measures should be placed to prevent and mitigate accidental radiological consequences.
9. Emergency preparedness and response for radiological incidents should be addressed.
10. Reduction of unregulated radiological risks should be ensured.

3. Nuclear waste management activities

Safe management of nuclear wastes is vital to ensure the sustainability of the nuclear industry in some countries and/or to end legacy practices in other countries. Historical waste management strategies relied on identifying two options to deal with the generated wastes, namely dilute and disperse and contain and confine [2, 11]. These options aimed at ensuring radiological protection for worker, public and the environment. Currently, nuclear wastes are managed according to contain and confine option. The objective of this option is to prevent radiological hazard by isolating the waste for sufficient periods that allow the decay and limit release of short- and long-lived radionuclides, respectively [2, 12, 13]. To reach this end point, the wastes should be subjected to volume reduction (pre-treatment, and treatment), conditioning, and disposal in engineered facility [1, 2, 12, 14–32]. **Figure 2** illustrates an example of radioactive waste management scheme, and all the activities performed within the scheme should be complementary. It should be noted that these activities are divided as predisposal and disposal activities, where transport, characterization and segregation, transportation, pre-treatment, treatment, conditioning, and storage are pre-disposal activities. Waste acceptance criteria (WAC) are applied at each facility to ensure integrated safe performance of the overall management schemes [27]. Each activity in this scheme should be authorized by the regulatory body according to national legislative system [28, 33]. Different technological options could be applied in each activity; the selection of any is bounded by different technical and nontechnical factors [2, 27]. Technical factors include the waste characteristics (chemical, physical, radiological, and biological), technology maturity, robustness, and flexibility, and site characteristics [2]. However, nontechnical factors include socio-economical impacts, legal framework, and financial and technical resources availability [2, 27].

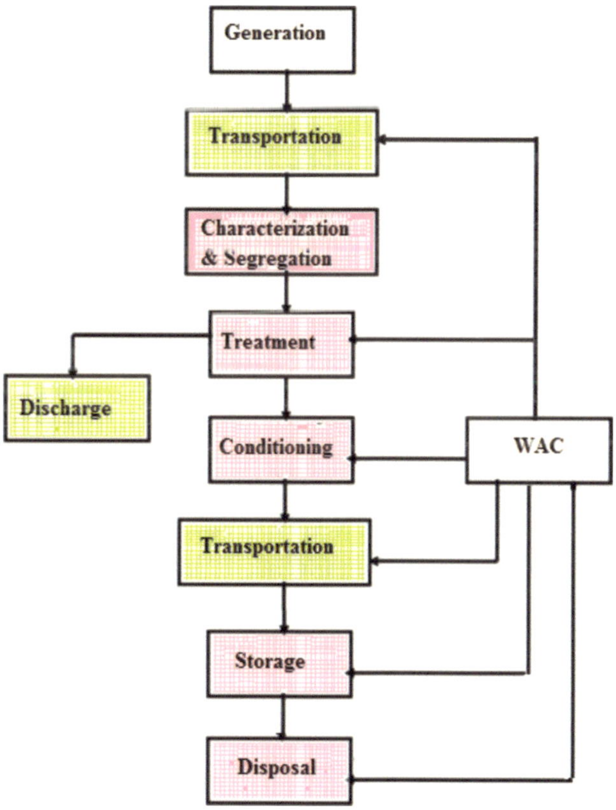

Figure 2. Radioactive waste management scheme.

4. Decision-making in nuclear industry

In the nuclear industry, decisions should be made based on sound scientific reasons that build confidence in the outputs of the industry [1, 33]. So, decisions are taken after ensuring effective improvement in safe design and operation of a facility/practice [34]. To support decision-making process, safety case (SC) is used to confirm the implementation of the required improvements. Safety case defined as a collection of all arguments that ensure achieving the highest level of quality in assessing radiological and nonradiological safety of the practice. These arguments include safety assessments, statement of confidence, and management system documentations. SC is prepared by dividing the studied system into subsystems that are analyzed and assessed, different terms are being used in this context, as follows [1, 4]:

Performance analysis: study of the system/subsystem/ process behavior and calculation of intermediate point of interest.

Safety analysis (*SA*): is directed to understand an overall system relevant to the protection against hazard.

Performance assessment (*PA*): determination of the performance acceptability that is conducted in comparison with certain design criteria or indicator.

Safety assessment (*SA*): aims at judging the overall system safety in comparison with regulatory safety limits, indicators and targets.

Indicators are characteristics that reflect possible impact of the system on the people/environment as a result of fault in a safety function or group of safety functions. Examples of safety functions for nuclear reactor and radioactive waste storage facility are listed in **Table 1** [9]. Based on the study types, indicators could be classified as performance or safety indicators. Performance indicators are usually used to assess the quality, reliability, or efficiency of the studied subsystems, whereas safety indicators are used to assess the performance of the overall system. **Table 2** lists the indicators classes and examples of these classes for radioactive waste disposal [34].

Hazard	Safety function in nuclear reactor	Safety function in radioactive waste storage facility
Criticality	Shut down and maintain shut down conditions of the reactor	Maintain subcriticality for fissile inventory
Thermal	Residual heat removal after shut down	Decay heat removal Control chemical process heat
Radiological	Confine radio-contaminants	Confine radioactive waste Shield radiation sources

Table 1. Safety functions in nuclear reactor and radioactive waste storage facility [9].

Indicator class	Indicators	Methods
Measurable	Spatial distribution of radionuclides in groundwater	Monitoring program in the site
Estimated	Lifetime of container	Derived from understanding the system
Calculated	Dose or risk	Modeling long-term evolution

Table 2. Examples of indicator classes in radioactive waste disposal system [34].

For radioactive waste disposal project, many decisions should be made at different milestones of the project lifecycle, that is, site selection, engineering design and construction, operation, closure, and post closure [35]. To support the decision-making process, the integration of assessment and analysis studies to produce radioactive waste disposal safety case should be conducted. **Figure 3** shows this integration, where the overall disposal system is divided into two main subsystems, near- and far-field that are analyzed and assessed. These subsystems are further divided into their main components, which are analyzed and assessed using relevant indicators. It should be noted that SC is a living document that is developed through the project lifecycle to reflect changes in the studied system, that is aging, operating experiences, and modifications. This is achieved by periodical safety review, where updated SC is submitted for revision on regular basis. IAEA identified the following requirements for safety case (SC) for radioactive waste disposal [36]:

1. SC is prepared by operator and reviewed and approved by regulator.
2. SC describes all safety relevant aspects of the site and facility, it includes SA and managerial control measures.
3. Adequate defense in depth is provided by applying several physical barriers and administrative procedures to ensure protection goal.
4. Site assessment should include present and natural evolution of the site and consider human plans and action in vicinity of the facility.
5. Facility development throughout its life cycle should preserve the identified safety function for the project.

Safety assessments represent a major part in safety case, as illustrated in **Figure 3**, safety assessment requirements for radioactive waste disposal facility include [36]:

1. SA should start at early stage of the disposal project and should be updated throughout the project lifecycle.
2. SA should conclude about compliance with safety objectives as required in the legislative framework.
3. Normal, anticipated and accidental conditions should be addressed in SA.

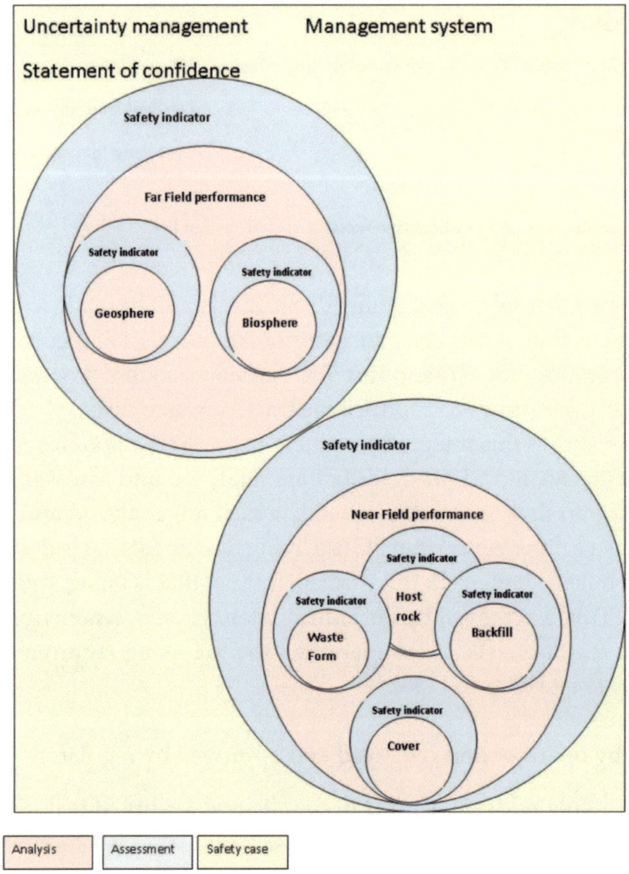

Figure 3. Integration of performance and safety assessment in radioactive waste disposal safety case.

4. Measures to control radiation risks that might arise under aforementioned conditions should be considered.

5. Radiation risks to individuals and population groups should be addressed for present and future generation.

6. SA could be conducted via deterministic and probabilistic approaches, and the scope of these analyses is determined according to the graded approach.

The outcomes of the decision-making process should assure [33, 34, 37]:

1. Maintenance of defense in depth and safety margins.

2. Consideration of good engineering and organizational practices.

3. Acknowledgement of the lesson learned from operational experience, research and development and state of the art.

4. Insurance of the integration of 3S concept.

5. Comply with relevant regulations.

5. Safety case development for radioactive disposal sites

Since the early establishment of radioactive/nuclear waste disposal practice, disposal was decided based on the output of performance and safety assessments. Both modeling and experimental assessments were carried out; **Table 3** lists the type of used assessment, studied subsystems, and performance indicators for three historical disposal practice [2, 38]. It should be noted that marine disposal is currently prohibited, hydro-fracture grouts are not used, whereas deep well injection license is renewed periodically [38].

Modeling radionuclide transport in near- and far-field is considered as the critical step in assessing the safety of the disposal practice. In modeling, processes that are important to the safety of the facility and site are linked together to predict the facility and site performances [39]. Modeling starts with system description, where information about important features, events, and processes is identified, then conceptual and mathematical models are developed. A typical scheme of a modeling process is illustrated in **Figure 4**. For an anticipated condition, conceptual model for disposal facility failure in a bathtub scenario is illustrated in **Figure 5**. The root cause of this scenario is the failure of the engineering barriers due to natural evolution of the system [39]. For each subsystem (waste form, engineering barriers, geo-sphere and bio-sphere), geometry, dimensionality, initial and boundary conditions, time dependence and relevant process are identified [40]. Then scenario consequences are determined using mathematical models. The level of mathematical model complexation is dependent on the stage of the disposal project. Within the design phase, three subphases are distinguished, namely,

Early disposal practice	Assessment type	Subsystems	Indicators
Marine	Modeling releases due to canister corrosion and package Monitoring release data	Near field: waste form, package	Radionuclide concentration
	Modeling radionuclide transport Monitoring transport	Far field: bottom sediment, benthic boundary layer, open oceans	Radionuclide concentration Doses
Hydro-fracture grout	Monitoring releases from experimental injection wells	Near field at monitoring wells	Radionuclide concentration
Deep well injection	Modeling transport Monitoring releases from experimental injection wells		

Table 3. Methods to assess the performance of historical disposal practice [38].

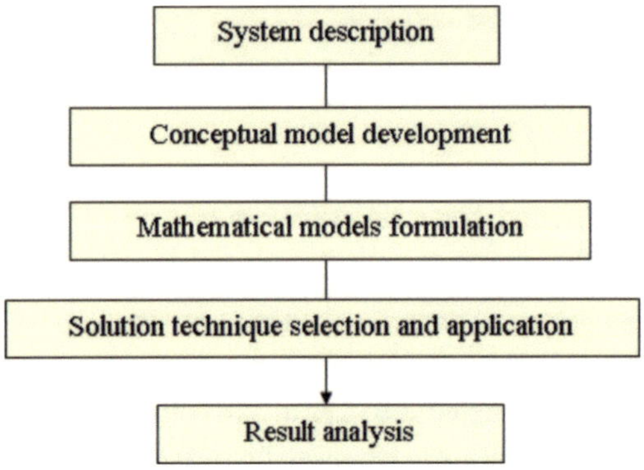

Figure 4. Steps in modeling process.

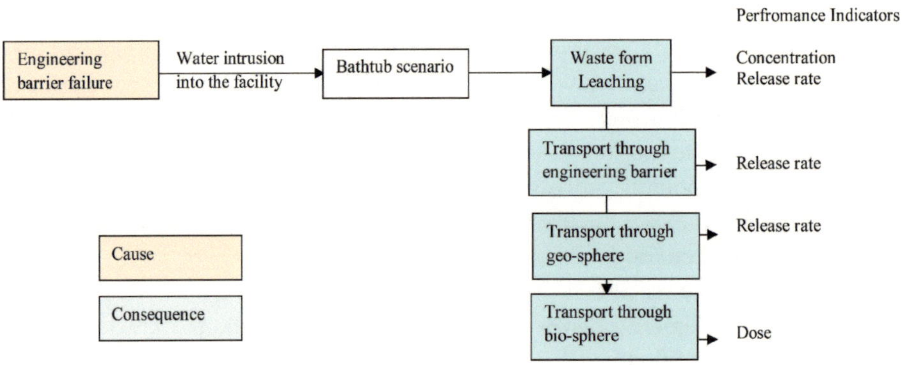

Figure 5. Conceptual model of disposal facility in a bathtub scenario.

conceptual design, basic design, and detailed design. The aim of the conceptual design is to select the disposal option; at this stage, data availability and its corresponding quality are respectively limited, so estimated radioactive waste inventory and generic site characteristics are used to model the system [2, 41]. As the project proceeds, modeling becomes more sophisticated where site-specific information and technical feasibility of engineering barrier materials become available. It should be noted that the development of the modeling efforts for each subsystem throughout the disposal lifecycle ensure the achievement of the sixth requirement for SA and third requirements for SC are addressed.

6. Conclusion

Nuclear engineering sciences emerged based on the integration of chemical, physical and engineering knowledge to serve the increased needs to optimize processes that utilizes radioactive, nuclear, and fusion reactions. The main objective of nuclear engineering is to ensure safe and sustainable production and application of nuclear/radioactive materials. This work introduced the role of nuclear safety in ensuring the sustainability of nuclear industry. It could be concluded that:

1. Safety case is used to support decision-making process, and it contains all the technical and managerial arguments that ensure safe operation of nuclear facility/practice under normal, anticipated, and accidental conditions.
2. The nature of safety case as a living document should be emphasized.
3. Safety case should assure the maintenance of the defense in depth and safety margins throughout the project life cycle.
4. Modeling is the core of safety case, and the level of modeling complicity should be proportional to the imposed risks and level of development of the project.

Author details

Rehab O. Abdel Rahman* and Hosam M. Saleh

*Address all correspondence to: alaarehab@yahoo.com

Hot Lab. Center, Atomic Energy Authority of Egypt, Cairo, Egypt

References

[1] Abdel Rahman RO, Kozak MW, Hung Y-T. Radioactive pollution and control, Ch 16. In: Hung YT, Wang LK, Shammas NK, editors. Handbook of Environment and Waste Management. Singapore: World Scientific Publishing Co; 2014. pp. 949-1027. DOI: 10.1142/9789814449175_0016

[2] Abdel Rahman RO, Rakhimov RZ, Rakhimova NR, Ojovan MI. Cementitious Materials for Nuclear Waste Immobilisation. New York: Wiley; 2014. ISBN: 9781118512005. http://eu.wiley.com/WileyCDA/WileyTitle/productCd-1118512006,subjectCd-CH50.html

[3] Abdel Rahman RO. Introduction to current trends in nuclear material research and technology, Ch 1. In: Abdel Rahman RO, Saleh HEM, editors. Nuclear Material Performance. Rijeka, Croatia: InTech; 2016. pp. 3-14. DOI: 10.5772/61411 http://www.intechopen.com/books/nuclear-material-performance/introductory-chapter-introduction-to-current-trends-in-nuclear-material-research-and-technology

[4] IAEA. Terminology Used in Nuclear Safety and Radiation Protection. 2007th ed. Vienna: International Atomic Energy Agency; 2007

[5] Toxicity, glossary. https://nanopartikel.info/en/glossary/234-toxicity [Accessed: 28 Feb, 2018]

[6] Medical definition of toxicity. https://www.medicinenet.com/script/main/art.asp?articlekey=34093 [Accessed: 28 Feb, 2018]

[7] OSHA, Appendix b to §1910.1200—Physical Hazard Criteria. https://www.osha.gov/dsg/hazcom/appendix_b.pdf [Accessed: 28 Feb, 2018]

[8] IAEA. Handbook of Nuclear Law: Implementing Legislation. Vienna: International Atomic Energy Agency; 2010

[9] IAEA. Fundamental Safety Principals, IAEA Safety Standard Series, SF.1. Vienna: International Atomic Energy Agency; 2006

[10] IAEA. Procedures for Conducting Probabilistic Safety Assessment for Non-reactor Nuclear Facilities. Vienna: International Atomic Energy Agency; 2002

[11] Abdel Rahman RO. Planning and implementation of radioactive waste management system. Ch 1. In: Abdel Rahman RO, editor. Radioactive Waste. Rijeka, Croatia: InTech; 2012. ISBN: 978-953-51-0551-0. http://www.intechopen.com/books/radioactive-waste/planning-and-implementation-of-radioactive-waste-management-system

[12] Abdel Rahman RO, El Kamash AM, Zaki AA, ElSourougy MR. Disposal: A last step towards an integrated waste management system in Egypt. In: International Conference on the Safety of Radioactive Waste Disposal; Tokyo, Japan, IAEA-CN-135/81. 2005. pp. 317-324

[13] Abdel Rahman RO, Ibrahim HA, Hanafy M, Abdel Monem NM. Assessment of synthetic zeolite NaA-X as sorbing barrier for strontium in a radioactive disposal facility. Chemical Engineering Journal. 2010;**157**:100-112

[14] Abdel Rahman RO, Ibrahium HA, Hung Y-T. Liquid radioactive wastes treatment: A review. Water. 2011;**3**:551-565

[15] Abdel Rahman RO, El-Kamash AM, Shehata FA, El Sourougy MR. Planning for a solid waste management quality assurance program in Egypt. Quality Assurance Journal. 2007;**11**:53-59

[16] Abdel Rahman RO. Design a quality control system for radioactive aqueous waste treatment facility. Quality Assurance Journal. 2009;**12**:31-39

[17] Saleh HM, Tawfik ME, Bayoumi TA. Chemical stability of seven years aged cement–PET composite waste form containing radioactive borate waste simulates. Journal of Nuclear Materials. 2011;**411**(1-3):185-192

[18] Bayoumi TA, Reda SM, Saleh HM. Assessment study for multi-barrier system used in radioactive borate waste isolation based on Monte Carlo simulations. Applied Radiation and Isotopes. 2012;**70**(1):99-102

[19] Eskander SB, Bayoumi TA, Saleh HM. Performance of aged cement-polymer composite immobilizing borate waste simulates during flooding scenarios. Journal of Nuclear Materials. 2012;**420**(1-3):175-181

[20] Eskander SB, Saleh HM. Cement mortar-degraded spinney waste composite as a matrix for immobilizing some low and intermediate level radioactive wastes: Consistency under frost attack. Journal of Nuclear Materials. 2012;**420**(1-3):491-496

[21] Saleh HM. Water hyacinth for phytoremediation of radioactive wastes simulate contaminated with cesium and cobalt radionuclides. Nuclear Engineering and Design. 2012;**242**:425-432

[22] Saleh HM, Eskander SB. Characterizations of mortar-degraded spinney waste composite nominated as solidifying agent for radwastes due to immersion processes. Journal of Nuclear Materials. 2012;**430**(1-3):106-113

[23] Eskander SB, Bayoumi TA, Saleh HM. Leaching behavior of cement-natural clay composite incorporating real spent radioactive liquid scintillator. Progress in Nuclear Energy. 2013;**67**:1-6

[24] Bayoumi TA, Saleh HM, Eskander SB. Solidification of hot real radioactive liquid scintillator waste using cement-clay composite. Monatshefte fur Chemie—Chemical Monthly. 2013;**144**(12):1751-1758

[25] Saleh HM. Stability of cemented dried water hyacinth used for biosorption of radionuclides under various circumstances. Journal of Nuclear Materials. 2014;**446**(1-3):124-133

[26] Saleh HM, Bayoumi TA, Mahmoud HH, Aglan RF. Uptake of cesium and cobalt radionuclides from simulated radioactive wastewater by Ludwigiastolonifera aquatic plant. Nuclear Engineering and Design. 2017;**315**:194-199

[27] Abdel Rahman RO, Metwally SS, El-Kamash AM. Life cycle of ion exchangers in nuclear industry: Application in radioactive wastes treatment and management of exhausted exchangers. In: Martínez LMT, Kharissova OV, Kharisov BI, editors. Handbook of Ecomaterials. Switzerland: Springer; 2018. DOI: 10.1007/978-3-319-48281-1_108-1 https://link.springer.com/referenceworkentry/10.1007/978-3-319-48281-1_108-1

[28] Abdel Rahman RO, El Kamash AM, Ali HF. Yung-Tse hung, overview on recent trends and developments in radioactive liquid waste treatment part 1: Sorption/ion exchange technique. International Journal of Environmental Engineering. 2011;**2**(1):1-16 http://serialsjournals.com/serialjournalmanager/pdf/1332219029.pdf

[29] El-Kamash AM, Mohamed RO, Nagy ME, Khalill MY. Modeling and validation of radionuclides releases from an engineered disposal facility. Radioactive Waste Management and Environmental Restoration. 2002;**22**(4):373-393

[30] Abdel Rahman RO, Zaki AA. Assessment of the leaching characteristics of incineration ashes in cement matrix. Chemical Engineering Journal. 2009;**155**:698-708

[31] Abdel Rahman RO, Zein DH, Abo Shadi H. Assessment of strontium immobilization in cement-bentonite matrices. Chemical Engineering Journal. 2013;**228**:772-780

[32] Abdel Rahman RO, Zein DH, Abo Shadi H. Cesium binding and leaching from single and binary contaminant cement-bentonite matrices. Chemical Engineering Journal. 2014;**245**:276-287

[33] IAEA. A Framework for an Integrated Risk Informed Decision Making Process, INSAG-25. Vienna: International Atomic Energy Agency; 2011

[34] IAEA. Safety Indicators for the Safety Assessment of Radioactive Waste Disposal, TEC-DOC1372. Vienna: International Atomic Energy Agency; 2003

[35] Abdel Rahman RO, El-Kamash AM, Zaki AA, Abdel-Raouf MW. Planning closure safety assessment for the Egyptian near surface disposal facility. In: International Conference on the Safety of Radioactive Waste Disposal; Tokyo, Japan, IAEA-CN-135/09. 2005. pp. 35-38

[36] IAEA. The safety case and safety assessment for the disposal of radioactive waste, SSG-23. Vienna: International Atomic Energy Agency; 2012

[37] IAEA. Safety Assessment for Facilities and Activities, GSR4. Vienna: International Atomic Energy Agency; 2016

[38] Abdel Rahman RO, Guskov A, Kozak M, Hung YT. Rcent evaluation of early radioactive disposal practice, Ch 8. In: Wang LK, Wang MS, Hung YT, Shammas NK, editors. Handbook of Environmental Engineering, Natural Resources and Control Processes. Vol. 17. Switzerland: Springer; 2016. pp. 371-400. DOI: 10.1007/978-3-319-26800-2_8 http://link.springer.com/chapter/10.1007/978-3-319-26800-2_8/fulltext.html

[39] Abdel Rahman RO, El-Kamash AM, Zaki AA. Modeling the long term leaching behavior of 137Cs, 60Co, and 152,154Eu radionuclides from cement-clay matrices. Hazardous Materials. 2007;**145**:372-380

[40] Abdel Rahman RO, Zaki AA. Comparative study of leaching conceptual models: Cs leaching from different ILW cement based matrices. Chemical Engineering Journal. 2011;**173**:722-736

[41] Abdel Rahman RO. Preliminary evaluation of the technical feasibility of using different soils in waste disposal cover system. Environmental Progress & Sustainable Energy. 2011;**30**(1):19-28

Section 2

Radiation Effects

Chapter 2

Radioisotope: Applications, Effects, and Occupational Protection

Sannappa Jadiyappa

Additional information is available at the end of the chapter

http://dx.doi.org/10.5772/intechopen.79161

Abstract

This chapter presents a brief introduction to radioisotopes, sources and types of radiation, applications, effects, and occupational protection. The natural and artificial sources of radiations are discussed with special reference to natural radioactive decay series and artificial radioisotopes. Applications have played significant role in improving the quality of human life. The application of radioisotopes in tracing, radiography, food preservation and sterilization, eradication of insects and pests, medical diagnosis and therapy, and new variety of crops in agricultural field is briefly described. Radiation interacts with matter to produce excitation and ionization of an atom or molecule; as a result physical and biological effects are produced. These effects and mechanisms are discussed. The dosimetric quantities used in radiological protection are described. Radiological protections and the control of occupational and medical exposures are briefly described.

Keywords: radioisotopes, sources of natural and artificial radiation, radiation dose, biological effect, dosimetric units, radiation protection, radiation application, environment

1. Introduction

Radiation and radioactivity existed long before life evolved on the earth and are *indispensable* parts of the environment. We are continuously exposed to natural and artificial radiations. In addition to these, some of the radionuclides such as polonium and radium are present in our bones; our muscle contains radiocarbon and radiopotassium, radon, thoron, and their progeny in our lungs, and they emit ionizing radiation. The radiation coming from the sun is due to the nuclear fusion; it is very essential for the existence of life on earth. Therefore we live in a natural radioactive world. All organisms including human beings on the earth are getting benefits from radiation in a direct way without realizing it. Therefore, without radiation life does not exist. Scientific understanding of radiation and radioactivity and their benefits and effects on humans, that's back almost century to the pioneering work of Roentgen (1895) and Becquerel (1896). Further investigation by M. Curie and P. Curie (1898) and Rutherford (1911) showed that radioactivity is exhibited by heavy elements such as uranium, thorium, and radium. The discovery of isotope was one of the results of work on the radioactive elements. The name "isotope" was first suggested by Soddy in 1913. The radioactive decay law was also proposed by him. More than thousand natural radioisotopes are present in our nature. At present more than 200 radioisotopes were produced from nuclear reactors and accelerators. The application of radioisotopes in medical, industry, and research field has served human civilization over a several decades. The radioisotopes have been a valuable gift to many braches of medicine and biology. Shorter half-lives of radioisotopes are used in medicine because they decay quickly and they are suitable for medical diagnosis and therapy. The World Health Organization (WHO) and International Atomic Energy Agency (IAEA) jointly coordinated a research program on radioactive tracers in cardiovascular diseases and searched for clues to this widespread health problem [1]. There are numerous applications of radioisotopes in medical fields; one of the revolutionized techniques is radioimmunoassay; this is used to detect and quantify minute levels of tissues components such as hormones, enzymes, or serum proteins by measuring the components ability to bind to an antibody or other proteins in competition with a standard amount of the same component that had been radioactivity tagged in the laboratory. For this technique, Rosalyn Sussman Yalow was awarded Nobel Prize in 1977 [2]. The precise dose is a life-and-death matter; therefore the IAEA has several program components to assist institutions in the members of the countries and aspect of radiation therapy and diagnosis. The IAEA in cooperation with the WHO offers on intercomparison service to check and improve accuracy of radiation dosimetry due to increase in the effectiveness of the radiotherapy [1]. The release of radioisotopes from nuclear fuel cycles, naturally occurring radioactive materials (NORM) from mining activity, mishandling of radioisotopes in industries and laboratories, and accidental release of radioactive materials could enter into the atmosphere. Therefore, it is necessary to require an urgent decision for protective actions. Therefore, the main objective is to focus on the applications and effects of radioisotopes and radiological protection.

2. Radioisotopes and radiation

The atom is the basic building block of matter. The concept of atoms and molecules was first introduced by John Dolton in 1811, and he proposed the atomic theory. The atom consists of positively charged nucleus and surrounded by a number of negatively charged electron, so that atom as a whole is electrically neutral. The electron had been discovered by J. J. Thomson in 1897. The nucleus consists of positive-charged proton and neutral-charged neutron referred as nucleons. The nucleus and proton were discovered by Rutherford in 1911, and neutron was discovered by James Chadwick in 1932. The number of proton present in the nucleus is called atomic number (Z), and total number of neutrons and protons present in the nucleus is called mass number (A). The atomic number of an element is the same, but different mass numbers are called isotope of an element. If the nucleus contains either excess of neutrons or protons, the force between these constituents will be unbalanced leading to unstable nucleus. An unstable nucleus will continuously vibrate and will attempt to reach stability by undergoing radioactive decay. The number of neutrons determines whether the nucleus is radioactive or not. The radioactive isotopes of an element are called radioisotopes; they are natural and artificially produced by nuclear reactors and accelerators. The discovery of radioisotope was one of the result works on the radioactive element. The way in which isotope arises in the radioactive element can be understood in terms of effects of radioactive decay on the atomic number Z and mass number A. In the year 1902, Rutherford and Soddy established that radioactivity is directly connected to the state of atomic nucleus.

The unstable nuclei of an element can undergo the variety of processes resulting in the emission of radiation in two forms, namely, radioactivity and nuclear reactions. In a radioactive decay, the nucleus spontaneously disintegrates to different species of nuclei or to a lower energy state of the same nucleus with the emission of alpha (α), beta (β), and gamma (γ) radiation is called radioactivity. The radioactivity was discovered by Henry Becquerel in 1896. Alpha, beta, and their ionizing property were discovered by Rutherford in 1899, and gamma was discovered by Villard in 1900. In nuclear reaction, the nucleus interacts with another particle or nucleus with subsequently emission of radiation as one of its final products. In some cases, the final product is also radioactive. The radiation emitted in both these processes may be electromagnetic (X-rays and γ-rays) or particle-like α, β, and neutrons. The nuclear reactions were discovered by Rutherford in 1917.

2.1. The type of emission of ionizing radiations

The ionizing radiations such as α, β, and γ except neutron are originated from unstable nuclei of an atom in an element undergoing radioactive decay.

2.1.1. Alpha radiation

Some naturally occurring heavy nuclei with atomic number $82 < Z < 92$ and artificially produced transuranic element $Z > 92$ decay by alpha emission, in which the parent nucleus loses both mass and charge. The alpha particle is emitted in preference to other light particles such as deuteron (^2H), tritium (^3H), and helium (^3He). Because energy must be released in order for decay to take place at all. The alpha particle has very stable and high binding energy, has tightly bound structure, and can be emitted spontaneously with positive energy in alpha decay, whereas ^2H, ^3H, and ^3He decay would require an input energy. The parent nucleus (Z, A) is transformed via

$$ {}_{Z}^{A}X \rightarrow {}_{Z-2}^{A-4}X + \alpha \quad (1) $$

It has less penetrating and high ionizing power.

2.1.2. Beta radiation

Beta particles are fast electron or positron; these are originated from weak interaction decay of a neutron or proton in nuclei, which contains an excess of the respective nucleon. In a neutron-rich nucleus, neutron can transform itself in to a proton by emission of beta particles and antineutrino. Similarly, in the nuclei with rich proton, it transforms into neutron by emission of neutrino and positron. These radiations are high penetrating and less ionizing power:

$$n \rightarrow p + e^- + \nu^- \quad (2)$$

Similarly in the nuclei with rich proton, the decay is

$$p \rightarrow n + e^+ + \nu \quad (3)$$

2.1.3. Gamma radiation

The emission of gamma rays is usually the most common mode of nuclear excitation and also occurs through internal conversion.

2.1.4. X-ray radiation

X-rays arises from the electron cloud surrounding the nucleus. They were discovered by Roentgen in 1895. X-rays are produced in X-ray tube by fast moving electron which is suddenly stopped by target.

2.1.5. Neutron radiation

It is a neutral particle that produces ionization indirectly by emission of γ-rays and charged particles when interacting with matter. These charged particles produce the ionization. It has more penetrating than gamma ray and can be stopped by thin concrete or paraffin barrier. They are produced by nuclear reaction and spontaneous fission in nuclear reactors. The characteristic emission of α, β, γ, and neutron sources is given in **Table 1** [3].

2.2. Classification of radiation

Depending on its effects on matter and its ability to ionize the matter, radiation is classified in two main categories: ionizing and nonionizing radiations.

2.2.1. Ionizing radiation

Radiation passing through the matter which breaks the bonds of atoms or molecules by removing the electron is called ionization radiation. It passes through the matter or living organisms, and it produces various effects.

Ionizing radiation is produced by radioactive decay, nuclear fission, and fusion, by extremely hot objects, and by particle accelerators. The emission of ionizing radiation is explained in Section 2.1. The ionizing radiation is again divided into two types: direct and indirect ionizing radiation.

2.2.1.1. Direct ionizing radiation

Directly ionizing radiation deposits energy in the medium through direct Coulomb interaction between the ionizing charged particles and orbital electrons of atoms in the medium, for example, α, β, protons, and heavy ions.

Source/isotope	Half-life	Energy (MeV)	
α			
^{241}Am	433 years	5.486	
^{210}Po	138 days	5.443	
^{242}Cm	163 days	5.305	
^{243}Am	7.4×10^3 years	6.113	
^{239}Pu	2.4×10^4 years	6.070	
β			
$^{3}_{1}H$	12.26 years	0.0186	
^{14}C	5730 years	0.156	
^{36}Cl	3.08×10^5 years	0.714	
^{63}Ni	92 years	0.067	
^{204}Tl	3.81 years	0.766	
γ			
^{60}Co	5.2 years	0.662	
^{137}Cs	30 years	1.277	
^{22}Na	2.6 years	1.173	
$^{60}_{27}C$	5.2 years	1.332	
X-rays			
^{41}Ca	8×10^5 years	3.690 keV	
^{44}Ti	48 years	4.508	
^{49}V	330 days	4.949	
^{55}Fe	2 k.6 years	5.895	
Source	**Half-life**	**Energy MeV**	**Yield × 10^6**
Neutron			
$^{239}Pu/Be$	24,000 years	5.14	65
$^{210}Po/Be$	138 days	5.30	73
$^{238}Pu/Be$	87.4 years	5.48	79
$^{241}Am/Be$	433 years	5.48	82

Table 1. Characteristics of some α, β, and γ emitters and neutron (sources).

2.2.1.2. *Indirect ionizing radiation*

Indirectly ionizing radiation deposits energy in the medium through a two-step process; in the first step, charged particles are released in the medium. In the second step, the released charged particles deposit energy to the medium through direct coulomb interaction with orbital electron of the atoms in the medium, for example, X-rays, photons, γ rays, and neutrons.

2.2.2. *Nonionizing radiation*

Nonionizing radiation is part of the electromagnetic radiation where there is insufficient energy to cause ionization. But it has sufficient energy only for excitation and not to produce ions when passing through matter [4]. Radiowaves, microwaves, infrared, ultraviolet, and visible radiation are the examples of nonionizing radiations. Nonionizing radiation is essential to life, but excessive exposures will cause biological effects.

3. Sources of natural and artificial radiation

There are two important sources of radiation: they are natural and man-made.

3.1. Natural background radiation

The radiation that exits all around us is called natural background radiation. All living organisms including man have been continuously exposed to ionizing radiations emitted from different sources, which always existed around us. The sources of natural radiation are cosmic rays and naturally occurring primordial radionuclides such as ^{238}U, ^{232}Th, ^{235}U, and their decay products as well as the singly occurring natural radionuclides like ^{40}K and ^{87}Rb, which are present in the earth crust, soil, rocks, building materials, ore, and water in the environment [5, 6]. Background radiation is a constant source of ionizing radiation present in the environment and emitted from a variety of sources. Natural radiations originated from three major sources: terrestrial, extraterrestrial, and internal (intake of natural radionuclides and their daughter product) sources of radiation.

3.1.1. Terrestrial sources of radiations

Terra means earth; the radiation originated from the earth crust is called terrestrial radiation. The primordial radionuclides (^{238}U, ^{232}Th, and ^{40}K) present in varying amounts in soil, rocks, water, and atmosphere are the sources of terrestrial radiation. The bulk of the natural radiation is mainly due to ^{40}K and ^{238}U, ^{232}Th, and their decay products [7]. Natural uranium consists of three isotopes ^{234}U, ^{235}U, and ^{238}U. ^{238}U is present in an abundance of 99.28% with a half-life of 4.5×10^9 years and ^{235}U in abundance 0.72% with a half-life of 0.7×10^9 years. Thorium is one of the important natural primordial radionuclides with a half-life of 1.4×10^9 years. It is about four times more abundant in nature than uranium. Average crustal abundance of ^{232}Th is 7.2 ppm [7]. All substances found in the terrestrial system contain variable amounts of ^{238}U and ^{232}Th; they undergo radioactive decay until they become stable isotopes. The two main important radioactive series are given in **Tables 2** and **3**.

The bulk of natural radiation comes from the primordial radionuclides such as ^{238}U, ^{235}U and ^{235}Th. They decay into other radioactive isotope as a part of radioactive series. These series are naturally occurring radioactive series, which have existed since the earth was formed. The nuclei in each series decay by emitting α, β and γ particles until stable (lead). These radioisotopes are chemically bound to minerals in rocks and soils and pose no biological hazards except radon, thoron and its progeny. Radon and thoron are noble radioactive gases, the higher concentrations of these gases and progenies are inhaled to produce lung cancer. According WHO and UNSCEAR, radon and their progeny are the second leading lung cancer after tobacco smoking.

http://dx.doi.org/10.5772/intechopen.79161

Parent nuclide	Half-life $T_{1/2}$	Decay mode (% branch)	Decay energy (MeV)	% Intensity	Daughter nuclide	γ-emission energy (keV)	% γ-emission intensity
^{238}U	4.5 × 10^{9} years	α (100)	4.198	79.0	^{234}Th	49.55	0.063
			4.151	20.9		113.50	0.0102
^{234}Th	24.10 days	B (100)	0.199	70.3	^{234}Pa	63.28	4.1
			0.104	19.2		92.37	2.4
			0.103	7.6		92.79	2.39
^{234}Pa	1.17 m	β (99.84)	2.269	98.2	^{234}U	1001.03	0.837
			1.224	1.007		766.38	0.294
		IT (0.16)	*		^{234}Pa	73.92	*
^{234}Pa	6.70 h	β (100)	0.642	19.4	^{234}U	131.30	0.029
			0.472	33.0		946.00	0.021
^{234}U	2.5 × 10^{5} years	α (100)	4.7746	71.38	^{230}Th	53.20	0.123
			4.7224	28.42		120.90	0.0342
^{230}Th	7.5 × 10^{4} years	α (100)	4.6870	76.3	^{226}Ra	67.672	0.373
			4.6205	23.4		143.872	0.0483
^{226}Ra	1600 years	α (100)	4.7843	94.45	^{222}Rn	186.21	3.59
			4.601	5.55		262.27	0.0050
^{222}Rn	3.8235 days	α (100)	5.4894	99.92	^{218}Po	511.00	0.076
^{218}Po	3.10 m	α (99.98)	6.0024	100.0	^{214}Pb	**	
		β (0.02)	*		^{218}At		
^{218}At	1.60 s	α (100)	6.0024	100.0	^{214}Bi	*	
^{214}Pb	26.8 m	β (100)	0.671	48.9	^{214}Bi	351.93	35.1
			0.728	42.2		295.22	18.2
			1.023	6.3		241.99	7.12
^{214}Bi	19.9 m	β (99.98)	3.272	18.2	^{214}Po	609.31	44.6
			1.542	17.8		1764.50	15.1
			1.507	17.02		1120.29	14.7
		α (0.02)	5.452	53.9	^{210}Tl	1238.11	5.78
			5.516	39.2		2204.21	4.98
^{214}Po	164.30 μs	α (100)	7.6868	99.99	^{210}Pb	799.7	0.0104
^{210}Tl	1.30 m	β (100)	4.209	30.0	^{210}Pb	*	
			1.863	24.0			
^{210}Pb	22.3 years	β (100)	0.0166	0.0631	^{210}Bi	46.54	4.25
			0.0631	16.0			
^{210}Bi	5.013 days	β (100)	1.1615	100	^{210}Po	**	
^{210}Po	138.376 days	α (100)	5.3043	99.99	^{206}Pb	803.10	0.00122
^{206}Pb	Stable end product						

**No gamma rays observed.

Table 2. Decay series of uranium (^{238}U) [8].

3.1.1.1. Uranium decay series

The decay series of uranium and the type of radiation and range of energy of decay products are shown in **Table 2** [8]. The important daughter product of uranium series is radon and its progenies. Radon is a naturally occurring radioactive gas. This was discovered by F.E. Dorn in 1900. It is found everywhere as part of our environment (i.e., in soil, water, and air). The ubiquitous radioactive gas is formed by radioactive decay of radium (^{226}Ra), which is the daughter product of uranium decay series (**Table 2**). The half-life of radon is 3.82 days; it decays by emission of alpha particle to form radon decay products or progeny, which are divided into short-lived and long-lived progeny. These are the significant contributor of natural radiation [9]. On the basis of the epidemiological studies, it has been established that the enhanced levels of indoor radon in dwellings can cause health hazards and may lead to serious diseases like lung cancer in human beings [5, 10].

3.1.1.2. Thorium decay series

The decay series of thorium and types of radiation with range of energies of decay products are as shown in **Table 3** [11]. The important daughter products in this series are thoron and its progenies. The thoron progeny has relatively long half-life than that of radon progeny; therefore thoron progeny would give a significant dose to the lungs [11–13]. The decay of thorium ^{232}Th leads to the subsequent formation of thoron (^{220}Rn), its half-life 55 seconds. It is more abundant than ^{238}U, but the short half-life of ^{220}Rn allows only a fraction to escape into the atmosphere. The ^{222}Rn is the one of the most significant isotope and it contributes significant dose to publics as ionizing radiation.

3.1.1.3. Potassium

Potassium is the most common of the naturally occurring non-series, singly occurring primordial radionuclides. Natural potassium comprises three isotopes ^{39}K (93.3%), ^{40}K (0.012%), and ^{41}K (6.7%). Among the three naturally occurring potassium isotopes, only ^{40}K is radioactive with a half-life of 1.28×10^9 years and having a specific activity of 31.4 Bqg^{-1} for natural potassium. ^{40}K decays through β-decay to stable ^{40}Ca 89% of the time. The remaining 10.72% of ^{40}K undergoes decay by electron capture to stable ^{40}Ar. This latter decay branch also emits a characteristic gamma ray at 1.461 MeV. This line is very useful to identify and quantify ^{40}K by gamma spectrometry [14].

Potassium is present in the earth crust with varying amounts and also present in almost all plant and animal tissues. Most of the potassium occurs in earth crust as minerals such as feldspar, orthoclase, muscovite, and biotite micas. Human beings require potassium to sustain their biological processes. A person who weighs 70 kg has about 140 g of potassium in his body which has activity of 4 kBq, most of which is located in the muscle. The absorbed dose per year is about 0.2 mSv in the bone. ^{40}K can be taken into the body by drinking water, eating food, or breathing air. Upon ingestion, ^{40}K then moves quickly from the gastrointestinal track into the bloodstream. The ^{40}K quickly enters the bloodstream and distributed to all organs and

Parent nuclide	Half-life $T_{1/2}$	Decay mode (% branch)	Decay energy (MeV)	% Intensity	Daughter nuclide	γ-emission energy (keV)	% γ-emission intensity
^{232}Th	1.4 × 10^{10} years	α (100)	4.0123	78.2	^{228}Ra	63.81	0.263
			3.9472	21.7		140.88	0.021
^{228}Ra	5.75 years	β (100)	0.0392	40.0	^{228}Ac	13.52	1.6
			0.0128	30.0		16.24	0.72
			0.0257	20.0		12.75	0.30
^{228}Ac	6.15 h	β (100)	1.158	29.9	228Th	911.20	25.8
			1.731	11.66		968.97	15.8
			2.069	8.0		338.32	11.27
^{228}Th	1.9116 years	α (100)	5.4232	72.2	^{224}Ra	84.373	1.22
			5.3404	27.2		215.98	0.254
^{224}Ra	3.66 days	α (100)	5.6854	94.92	^{220}Rn	240.99	4.10
			5.4486	5.06		292.70	0.0062
^{220}Rn	55.6 s	α (100)	6.2881	99.87	^{216}Po	549.76	0.114
^{216}Po	0.145 s	α (100)	6.7783	100	^{212}Pb	804.9	0.0019
^{212}Pb	10.64 h	β (100)	0.335	82.5	^{212}Bi	238.63	43.3
			0.574	12.3		300.09	3.28
			0.159	5.17		115.18	0.592
^{212}Bi	60.55 m	β (64.06)	2.248	86.57	^{212}Po	727.33	6.58
			1.521	6.81		1620.50	1.49
			0.627	2.92		785.37	1.102
		α (35.94)	6.0508	69.19	^{208}Tl	39.86	1.06
			6.0899	27.12		288.20	0.337
			5.7675	1.78		452.98	0.363
^{212}Po	0.299 μs	α (100)	8.7849	100.0	^{208}Pb	**	
^{208}Tl	3.053 m	β (100)	1.796	48.7	^{208}Pb	2614.53	35.64
			1.286	24.5		583.19	30.4
			1.519	21.8		510.77	8.13
^{208}Pb	Stable end product						

**No gamma rays observed.

Table 3. Decay series of thorium (^{232}Th) [8].

tissues. Each year, this isotope delivers doses of about 18 millirem (mrem) to soft tissues of the body and 14 mrem to the bone. ^{40}K can present both external and internal health hazards [9].

3.1.2. *Extraterrestrial radiations (cosmic radiations)*

The extraterrestrial radiations or cosmic radiations are high energetic radiations or subatomic particles, mainly originated from the sun, stars, collapsed stars (such as neutron stars), quasars, and the hot galactic and intergalactic plasma. The earth and all living things on it are constantly bombarded by these radiations from space. The galactic cosmic radiation coming at the upper atmosphere is made up of about 98% baryons and 2% electrons [15]. Cosmic ray radiation consists of 85% protons, 14% alpha particles (helium ions), and about 1% nuclei of atomic number between 4 and 26. The cosmic ray particles incident on the earth's atmosphere are the mixture of charged particles such as electrons, protons, α-particle, and a just detectable amount of heavier nuclei. These radiations have extremely high energies that vary from 10^2 MeV to more than 10^{14} MeV [16]. The cosmic ray shower (typically beta and gamma radiation) is obtained due to interaction of cosmogenic charged particles with the earth's atmosphere and magnetic field. The dose from cosmic radiation varies in different parts of the world due to differences in elevation and to the effects of the earth's magnetic field. The cosmic radiations are much more intense in the upper troposphere. Cosmic radiation dose increases with altitude; at 2.5 km, it is about 0.55 $mSv.y^{-1}$, and during their flights, airline crews typically get an extra dose on the order of 2.2 $mSv.y^{-1}$. Therefore, the annual effective doses from cosmic ray radiation around the world are estimated to range between 0.26 and 2.00 $mSv.y^{-1}$ [17].

The cosmic radiation interacts with the atoms and molecules in the atmosphere (mainly nitrogen and oxygen), are the dominant mechanism of interaction resulting in a cascade of interactions, and produces a number of secondary radiation in the form of protons, neutrons, and charged/uncharged pions of various energies which intern produce a various radioisotopes such as ^{3}H, ^{7}Be, ^{14}C, ^{22}N, ^{32}P, ^{35}S, and ^{35}Cl through nuclear reactions known as cosmogonic radionuclides [18]. These cosmogonic nuclides eventually reach the earth's surface and can be incorporated into living organisms and also contribute to the natural radiation exposures. The equivalent annual effective dose from cosmogonic radionuclides is estimated to be 12 μSv for ^{14}C, 0.15 μSv for ^{22}Na, 0.01 μSv for ^{3}H, and 0.03 μSv for ^{7}Be, and the most significant of these is due to ^{14}C [15].

3.1.3. *Internal sources of radiation*

The radioactive material gets inside the body by eating, drinking, breathing, or injection (from certain medical procedures and radon/thoron and their progeny). The alpha and beta radiation emitted by these radioactive materials poses serious health threat if significant quantities are inhaled or injected.

3.2. Artificial or man-made radiation

In addition to natural background radiation, human beings are exposed to man-made radiation obtained from nuclear installations, nuclear explosions, nuclear fuel cycle, radioactive

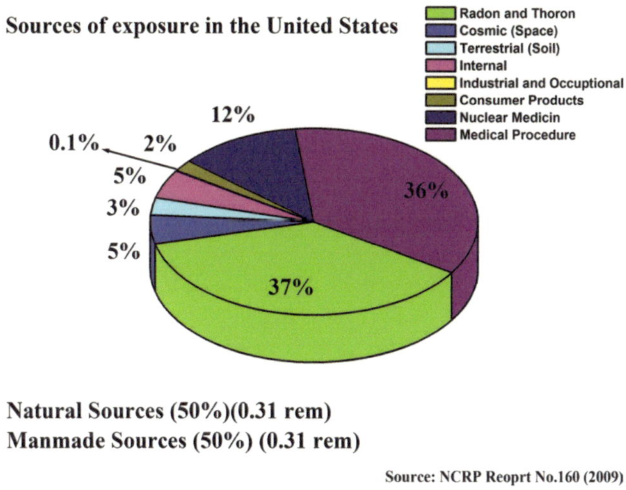

Figure 1. Sources of radiation exposure in the United States.

waste releases from nuclear reactor operations, and accidents and other industrial, medical, and agricultural uses of radioisotopes. The most significant sources of exposure, which gives the largest contribution to the public is from medical diagnostic X-rays, nuclear medicine, and nuclear therapy. This is also generated from consumer products such as combustible fluids (gas and coal), TV, luminous watches and dials, and electron tubes. The public are exposed to the radiation from the nuclear fuel cycle, which includes the entire sequence from mining and milling of uranium, the actual production of power at a nuclear power plant, and residual fall-out from nuclear weapon testing and accident. The public are not exposed to all the sources of radiation, for example, patients who are treated with the medical irradiation or the workers of nuclear industry may receive higher radiation exposure than the public [19]. The sources of radiation exposure in the United States were given in **Figure 1**.

4. Applications of radioisotope

The applications of radioisotopes have played a significant role in improving the quality of life of human beings. The whole world is aware of the benefits of the radiation, but the phobia of nuclear weapons on Hiroshima and Nagasaki (August 6 and 9, 1945) and the nuclear accidents occurred in Chernobyl in Russia (April 25–26, 1986) and Fukushima in Japan (March 2011) was so deep in the mind of the common man that we can still struggle to come out of it. Major problems arrived by workers in nuclear fields are due to lack of legalization, shortage of resources, and knowledge about nuclear society and safe guards. To minimize these problems the international organization such as International Atomic Energy Agency (IAEA) and International Commission on Radiation Protection (ICRP), identify requirements and provide the infrastructure that can support nuclear technology [20].

4.1. Radiotracer (radioisotopes)

Radiotracers are widely used in medicine, agriculture, industry, and fundamental research. Radiotracer is a radioactive isotope; it adds to nonradioactive element or compound to study the dynamical behavior of various physical, chemical, and biological changes of system to be traced by the radiation that it emits. The tracer principle was introduced by George de Hevesy in 1940 for which he was awarded the Nobel prize.

4.1.1. Radioisotope production

The sustainability of radioisotope production is one of the critical areas that receive great attention. There are more than 160 different radioisotopes that are used regularly in different fields; these isotopes are produced either in a medium or in high-flux research reactors or particle accelerators (low or medium energy) [21]. Some of the radioisotopes produced by the reactor and particle accelerators and their applications are given in **Table 4**.

Reactor radioisotope	Half-life	Applications
Radioisotopes produced by reactors		
Bismuth-213	45.59 min	It is an alpha emitter (8.4 MeV). Used for cancer treatment, e.g., in the targeted alpha therapy (TAT)
Cesium-131	9.7 days	It emits photon radiation in the X-ray range (29.5–33.5 keV). Used in brachytherapy of malignant tumors
Cesium-137	30 years	Used in medical devices (sterilization) and gauges (661.64 keV)
Chromium-51	28 days	Used in Diagnosis of gastrointestinal bleeding and to label platelets (320 keV)
Cobalt-60	5.27 years	Used for controlling the cancerous growth of cells (1173.2 keV)
Dysprosium-165	2 h	Used for synovectomy treatment of arthritis (95 keV)
Erbium-169	9.4 days	Used for relieving arthritis pain in synovial joints (8 keV)
Holmium-166	26 h	Diagnosis and treatment of liver tumors (81 keV)
Iodine-125	60 days	Used in cancer brachytherapy and radioimmunoassay (35 keV)
Iodine-131	8 days	Widely used in treating thyroid cancer and in imaging the thyroid, diagnosis, and renal blood flows (284 keV)
Iridium-192	74 days	Used as an internal radiotherapy source for cancer treatment. Strong beta emitter for high-dose rate brachytherapy (317 keV)
Iron-59	46 days	Used in studies of iron metabolism in the spleen (1095 keV)
Lead-212	10.6 h	Used in TAT for cancers (239 keV)
Molybdenum-99	66 h	Used as the parent in a generator to produce technetium-99 m (740 keV)
Palladium-103	17 days	Used to make brachytherapy permanent implant seeds for early-stage prostate cancer. Emits soft X-rays (362 keV)
Potassium-42	12.36 h	Used for potassium distribution in bodily fluids and to locate brain tumors (1524 keV)
Radium-223	11.4 days	Used to treat prostate cancers that have spread to the bones

Reactor radioisotope	Half-life	Applications
Rhenium-186	3.71 days	Used for therapeutic purpose to relief pain in bone cancer. Beta emitter with weak gamma for imaging (137 keV)
Samarium-153	47 h	Effective in relieving the pain of secondary cancers lodged in the bone, sold as Quadra met. Beta emitter (103 keV)
Selenium-75	120 days	Used to study the production of digestive enzymes (265 keV)
Sodium-24	15 h	Used for studies of electrolytes within the body (2754 keV)
Ytterbium-169	32 days	Used for cerebrospinal fluid studies in the brain (63 keV)
Radioisotopes produced by accelerators		
Cobalt-57	272 days	Used as a marker to estimate organ size and for in vitro diagnostic kits (122 keV)
Copper-64	13 h	Used for PET imaging studies of tumors and also cancer therapy (511 keV)
Copper-67	2.6 days	Beta emitter, used in therapy
Fluorine-18	110 min	Used as fluorothymidine (FLT)
Gallium-67	78 h	Used for tumor imaging and locating inflammatory lesions (infections)
Indium-111	2.8 days	Brain studies, infection, and colon transit studies
Iodine-123	13 h	Used for diagnosis of thyroid function
Rubidium-82	1.26 min	Convenient PET agent in myocardial perfusion imaging
Strontium-82	25 days	Used as the parent in a generator to produce Rb-82
Thallium-201	73 h	Used for location of low-grade lymphomas

Table 4. Some of the radioisotopes produced by the reactor and particle accelerators and their applications.

4.2. Medicine

Nowadays radiotracer has become an indispensable and sophisticated diagnostic tool in medicine and radiotherapy purposes.

4.2.1. Diagnostic purpose

The most common radioactivity isotope used in radioactive tracer is technetium (^{99}Tc). Tumors in the brain are located by injecting intravenously ^{99}Tc and then scanning the head with suitable scanners.

^{131}I and most recently ^{132}I and ^{123}I are used to study malfunctioning thyroid glands. Kidney function is also studied using compound containing ^{131}I. ^{33}P is used in DNA sequencing. Tritium (^{3}H) is frequently used as a tracer in biochemical studies. ^{14}C has been used extensively to trace the progress of organic molecule through metabolic pathways.

A most recent development is positron emission tomography (PET), which is a more precise and accurate technique for locating tumors in the body. A positron emitting radionuclide (e.g., ^{13}N, ^{15}O, ^{18}F, etc.) is injected to the patient, and it accumulates in the target tissue. As it emits positron which promptly combines with nearby electrons, it results in the simultaneous emission of two γ-rays in opposite directions. These γ-rays are detected by a PET camera and give precise indication of their origin, that is, depth also. This technique is also used in cardiac and brain imaging.

Compound X-ray tomography or CT scans. The radioactive tracer produces gamma rays or single photons that a gamma camera detects. Emissions come from different angles, and a computer uses them to produce an image. CT scan targets specific area of the body, like the neck or chest, or a specific organ, like the thyroid [22].

4.2.2. Therapeutic

The most common therapeutic use of radioisotopes is ^{60}Co, used in treatment of cancer. Sometimes wires or sealed needles containing radioactive isotope such as ^{192}Ir or ^{125}I are directly placed into the cancerous tissue. The radiations from the radioisotopes attack the tumor as long as needle/wire is in place. When the treatment is complete, these are removed. This technique is frequently used to treat mouth, breast, lung, and uterine cancer. ^{131}I is used to treat thyroid for cancers and other abnormal conditions of thyroid. ^{32}P is used to treat excess of red blood cells produced in the bone marrow.

4.3. Agricultural research

Development of high yielding varieties of plants, oil seeds, and other economically important crops and protection of plant against the insects are the thrust area of agricultural research.

4.3.1. New varieties of crops

The irradiated seeds of wheat, rice, maize, cotton, etc., are undergoing profound genetic changes in order to improve crop varieties and mutation breeding. These varieties of crops are more disease resistant and have high yields. Several countries all over the world produce new variety of crops from radiation-induced mutants [23].

4.3.2. Eradication of insect and pests

The best technique for the control of insects and pests is sterile insect technique (SIT). Irradiation is used to sterilize mass-reared insects so that, while they remain sexually competitive, they cannot produce offspring. As a result, it enhances the crop production and preservation of natural resources.

4.4. Food preservation and sterilization

As per WHO reports, about 25–35% of world food production is susceptible to the attack by pests, insects, bacteria, and fungi causing a great loss of the economy of the country. Food irradiation has more advantages than conventional methods. All types of radiations are not recommended for food irradiation; only three types of radiation are recommended by CODEX general standard for food irradiation which are ^{60}Co or ^{137}Cs, X-rays, or electron beams from particle accelerators [24]. The food products are exposed to γ-radiations from the intense controlled sources to kills pests, bacteria, insects, and parasites and extends shelf-life but also reduces the food's nutritional value somewhat by destroying vitamins A, B_1 (thiamin), C, and E. No radiation remains in the food after treatment.

Depending on the radiation dose and its application, radiations are classified into three categories: they are low dose (<1 kGy), medium dose (1–10 kGy), and high dose (>10 kGy) [24].

4.4.1. Low-dose applications

4.4.1.1. Sprout inhibition in bulbs and tubers

Irradiated potato can be stored at higher temperature of around 15°C. This not only conserves energy but also prevents sweetening of potato, commonly occurring at low temperatures. It gives advantage to the manufacturers of chips as low-sugar potato gives desired lighter color to fries and chips.

4.4.1.2. Delayed ripening of fruits

Irradiated fruits (all kinds of mangoes) of these at hard mature pre-climacteric stage at 0.25–0.75 kGy delay the ripening process by about 7 days, thus improving shelf-life. These doses are also effective in destroying quarantine pests. Irradiated fresh fruits can be stored for longer duration, sometimes up to 30 days at 12–14°C and in modified atmospheres.

4.4.2. Medium-dose applications

Under ice, sea food such as fish and prawns, fish-like Bombay duck, pomfret, Indian salmon, mackerel, and shrimp can be stored for about 7–10 days. Studies have demonstrated that irradiation at 1–3 kGy followed by storage at melting ice temperatures increases its shelf-life nearly threefold.

Meat and meat products including poultry have a shelf-life of about a week at 0–3°C, which could be extended up to 4 weeks by applying a dose of 2–5 kGy, which inactivates spoilage bacteria. Radiation treatment has been employed to enhance the shelf-life of intermediate moisture meat products.

4.4.3. High-dose applications

While transporting the spices, due to inadequate handling and processing conditions, spices get contaminated with insect eggs and microbial pathogens. When incorporated into semi-processed or processed foods, particularly, after cooking, the microbes, both spoilers and pathogens, in spices can outgrow causing spoilage and posing risk to consumers. Many of the spices develop insect infestation during storage, and unscrupulous traders convert them into spice powders. A dose of 10 kGy brings about near sterility or commercial sterility while retaining the natural characteristics of spices.

Irradiation at higher doses can also be employed for total sterilization of diets for immuno-compromised patients, adventure sports, military, and astronauts.

4.5. Industry and civil engineering

Radioisotopes are commonly used in industry for checking blocked water pipes and detecting leakage in oil pipes. For example, small quantity of radioactive ^{24}Na is placed in a small enclosed ball and is allowed to move in pipe with water. The moving ball containing radioisotope is monitored with a detector. If the movement of ball stops, it indicates the blocked pipe. Similarly, radioisotope ^{24}Na is mixed with oil flowing in an underground pipe. With radiation detector, the radioactivity over the pipe is monitored. If there is a leakage place, the radiation detector will show large activity at that particular place. Radioisotopes are also used to monitor fluid flow and filtration, detect leaks, and gauge engine wear and corrosion of process in equipment.

Radioactive materials are used to inspect metal parts and the integrity of welds across a range of industries. The titanium capsule is a radioactive isotope which is placed on one side of the object being screened, and some photographic film is placed on the other side. The gamma rays pass through the object and create an image on the film. Gamma rays show flaws in metal castings or welded joints. The technique allows critical components to be inspected for internal defects without damage. Radiotracer is also used to inspect for internal defect without damage.

In industries, the production methods need to be constantly monitored in order to check the quality of products and to control the production process. The monitoring is carried out by quality control devices using the unique properties of radiation; such devices are called nuclear gauges. They are more useful in extreme temperature, harmful chemical process, molten glass, and metals. The gauges are also used to measure the thickness of sheet materials, including metals, textiles, paper, and plastic production.

5. Effects of exposure to radioisotope

Radiation passing through the material breaks the bonds by removing the electron of an atom or molecules; this induces physical, chemical, and biological changes. Ionizing radiation focuses large amount of energy into a highly localized areas of irradiated materials. Damage is caused by the interaction of this energy with nuclei or orbiting electrons. The material structure may be modified through this energy interaction; as a result the mechanical property of bulk material changes.

5.1. Radiation effects on metals

Radiation creates a point defect in metals; this had been recognized by Wigner in 1946. The radiation effects on metals depend on type and duration of the radiation. Ionizing radiation can affect the metal in two ways, (1) lattice atoms are removed from their regular lattice sites, that is, displacement damage production and (2) chemical composition of the target can be changed by ion implantation or transmutation.

Neutron-irradiated metals at room temperature show increase in electrical and thermal resistance, hardness, and tensile strength and higher yield strength along with decrease ductility in metals [25]. At higher temperature it is found that the strength and ductility return to the same values as before irradiation. A metal under stress at higher temperature exhibits the phenomenon of creep, that is, the gradual increase in strain with time. The thermal neutrons have less significant effect on the mechanical properties of metals. They can be captured by nuclei of irradiated material which will become radioactive.

5.2. The radiation effects on nonmetals

Radiation causes the viscosity of oil and grease to increase as gummy, tar-like polymers are formed. Radiation causes soap-oil-type greases to become more fluid. Plastics undergo drastic changes when exposed to radiation. The rubber may become harder or softer depending on its types. Concrete under radiation exposure heats up. This drives the water out of its internal structure. Swelling, cracking, and spalling result [25].

5.2.1. Radiation effects on polymers

Ionizing radiation can alter the molecular structure and macroscopic properties of the polymer. The polymers are exposed to the radiation; as a result excitation and ionization of target/molecules are produced. These processes in the target molecules lead to breaking of original bonds, production of ionized and excited species, bond rearrangement, chain scission, radical formation, etc. All these processes are responsible for the modification of chemical, electrical, mechanical, and optical properties of polymers leading to their applications in different scientific and technological fields [25].

5.2.2. Cross-linking

During radiation polymerization, the interaction takes place between two free radical monomers which combine to form intermolecular bond leading to three-dimensional network of cross-linked high molecular polymer. These cross-linked polymers show high thermal resistance and strong mechanical strength [26].

5.2.3. Radiation grafting

Grafting is a method wherein monomers are covalently bonded (modified) onto the polymer chain. Irradiation on polymers by γ-radiation is useful for the functionalization of surfaces with stimuli–response polymers. This method involves the formation of free radical sites near the surface of polymers on to the polymeric backbone as a result of irradiation. Hence microenvironment suitable for the reaction among monomer or polymer and the active site is formed, leading to propagation to form side chain grafts. Radiation grafting changes the surface of polymeric materials by chemically bonding polar or nonpolar monomers having functional groups such as –COOH, -OR, $-NH_2$, $-SO_3H$, –R, and their derivatives, to affect surface properties. The radiation-induced grafting is used in variety of applications such as biomedical, environmental, and industrial uses [26–28]. The radiation grafting can be performed by two major methods: pre-irradiation technique and mutual or simultaneous method [29].

5.2.4. Degradation

Radiation-induced degradation technology is a new application to develop viscose, pulp, paper, food preservation, pharmaceutical production, and natural bioactive agent industries. Controlling the degree of degradation of polymers in industries is very important. Irradiation of polymers induces molecular chain branching, cross-linking, and molecular degradation or scissioning. Chain branching increases the molecular weight of the polymer. Cross-linking forms the insoluble three-dimensional polymer network, while degradation or scissioning causes a reduction of initial molecular weight [30, 31]. The polymer irradiated in air by solar radiation results in the formation of free radicals and can also react with oxygen, giving rise to oxidative degradation. All these molecular modifications can modify the properties of polymers. The study of degradation of polymer is important in using polymeric materials in radioactive environments such as in nuclear power plants, space, or the sterilization of polymeric medical disposals or food plastic packaging [32]. The splitting of polymeric macromolecules to form free radicals is employed for synthesizing modified polymers. At the same time, polymer degradation may often be considered as an undesirable side reaction occurring during the chemical transformation, fabrication, and usage of polymers.

5.3. Biological effects of radiations

The harmful effects that are produced in human beings who are exposed to radiations are called health effects. The result of all the physical interaction processes between incident radiation and the tissue of a cell is a trail of ionized atoms and molecules. The radiation is directly interacting with sensitive critical sites of the tissue (DNA) to produce damage by breaking chemical bonds. The chemically active free radicals are indirectly produced by interaction of primary radiation with DNA of the tissue. Both direct and indirect damages produced in DNA by radiation are shown in **Figure 2**.

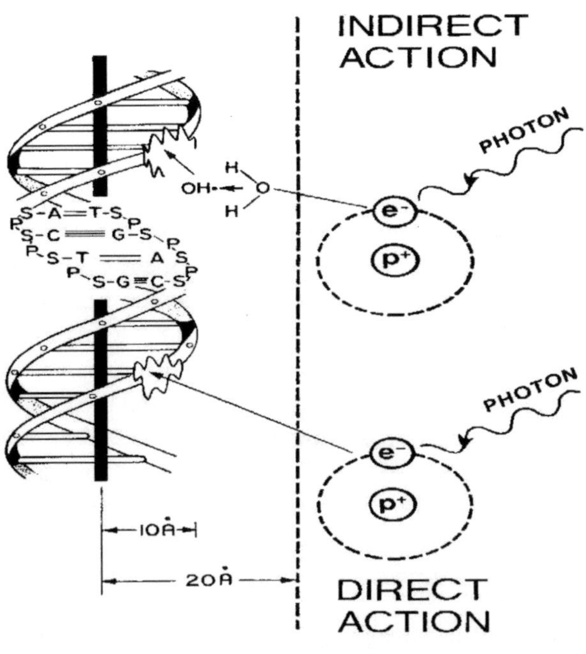

Figure 2. The mechanism by which damage occurs in the cell by direct and indirect action of radiation.

5.3.1. Direct action

Radiation attacks DNA molecule directly (**Figure 2**); as a result the ionization is produced and the bond is disrupted within a few nanometers of the DNA molecule.

5.3.2. Indirect action

Indirect action is due to the chemical radicals which are produced by radiation that interact with water molecule; it comprises about 80% of tissue. Free radicals are important since they can diffuse far enough to reach and induce chemical changes at critical sites in biological structures. The chemical damage produced by the breaking of DNA by the action of free radicals. The formation and action are as follows:

Ionization of a water molecule produces a free electron and a positively charged molecule:

$$H_2O + Radiation \rightarrow H_2O^+ + e^- \tag{4}$$

The released electron is most likely to be captured by another water molecule converting it into a negative ion:

$$H_2O + e^- \rightarrow H_2O^- \tag{5}$$

Both these ions are unstable and dissociate which are as follows:

$$H_2O^+ \rightarrow H^+ + OH^0 \tag{6}$$

$$H_2O \rightarrow H^0 + OH^- \tag{7}$$

Formation of free radicals is denoted by OH^0 and H^0. These free radicals interact with organic biomolecules (RH) again to produce organic free radicals denoted by R^0:

$$RH + OH^0 \rightarrow R^0 + H_2O \tag{8}$$

$$RH + H^0 \rightarrow R^0 + H_2 \tag{9}$$

These free radicals interact with DNA to produce the damage. DNA is made up of double-helix structure; if the radiation/free radical breaks only one strands, it is easily repaired by opposite strand as a template. If double strand breaks the repair of the cell is not possible; as a result mutations or changes in DNA code this leads to a cell death or cancer. To a certain extent, these molecules are repaired by natural biological processes, and this ability to self-heal or self-repair depends on the extent of damage. The biological effect of radiation on living cell may result in three outcomes:

1. Death of the cells
2. Impairment in the natural functioning of cell leading to somatic effects such as cancer
3. A permanent alteration of the cell which is transmitted to later generation, that is, genetic effect.

Oxygen effect is another effect produced by organic free radicals. The amplification of the Chemical action of free radicals due to the presence of oxygen in tissue is called oxygen effect.

It has consequences that irradiated cell have a lower chance of survival in tissue rich in oxygen than in tissue less rich in oxygen.

Biological effects of radiation are broadly classified into deterministic effect and stochastic effects.

5.3.3. Deterministic effect

These effects of damage from the radiation can be long term or short term. The large amount of radiation which is exposed to short interval of time is called acute radiation effects. Small amount of radiation dose exposed to longer period is called delayed effect or chronic effect. Deterministic effects are severe, if dose exceeds a threshold level (500 mSv). The severity of these effects in an exposed individual increases with the dose above the threshold as shown in **Figure 3(b)**. The acute effect above the threshold at different time intervals is given **Table 5**.

5.3.4. Stochastic effect

These effects are associated with long-term low-level exposure. They have no apparent threshold. The risk from the exposure increases with increasing the dose, but the severity of the effect is independent of the dose as shown in **Figure 3(a)**. In stochastic effect (**Figure 3(a)**) the effect rate increases with increase in the dose rate. But in deterministic effect (**Figure 3(b)**), there is apparent threshold, thereafter the effect rate increases rapidly with increase in the dose.

With the chronic exposure, there is a delay between the exposure and observed health effects. These effects include cancer and other health outcomes such as benign tumor, cataracts, and potentially harmful genetic effects. Cancer and genetic effects are recognized as stochastic effects.

5.3.5. Genetic effect

These effects are not immediate. They are produced in the future generation. The experimental evidences from animal studies show that the radiation can cause genetic effects, but the studies of the survivors of Hiroshima and Nagasaki gives no indication of these effects on human beings [33].

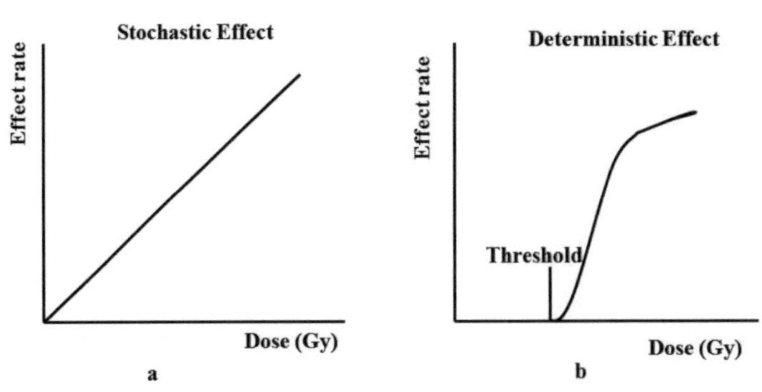

Figure 3. Deterministic effect and stochastic effects of radiation.

Effective dose	Time of exposure	Effects
<1 Sv	Weeks	Decrease white blood cell levels
	Short period	Recovery
Around 2 Sv	1 h	Nausea, headache, or vomiting
	Within a week	50% of lymphocyte decreases and 50% thrombocyte level decreases
	3–4 weeks	Quick recovery is possible
3 Sv	Hours to days	Radiation sickness: nausea, vomiting, fatigue, loss of appetite, infections, infections, dehydration, and hair loss
	After few days	Patient may feel better
	Within a week or months	Good recovery
4 Sv	Few weeks	Damage to the mucous membrane of the intestines and bone marrow tissue
5–6 Sv	—	A strong possibility of death
>6 Sv	—	The chance of surviving longer than a few weeks is slim
>10 Sv	Within 2 weeks	The mucous membrane of the intestines will be damaged beyond repair causing death
50 Sv	Within hours	The central nervous system will be damaged, loss of consciousness
	Within a day	Death

Table 5. Deterministic effect with different doses [37].

There is a considerable uncertainty about the low dose is beneficial or harmful. In the published literature, data and reports regarding health effects of low doses are two classes of thoughts. The first thought favors in linear no-threshold (LNT) hypothesis adopted by major scientific, official, and governmental organizations such as ICRP, NCRP, NAS-NRC, WHO, and UNSCEAR for risk assessment and states that the low radiation is harmful. The other school of thought believes in the beneficial features (hormesis hypothesis) of such a low-level exposure. According to this hypothesis, very low dose of radiation is beneficial, that is, to stimulate repair mechanism and induce activity of DNA region (UNSCEAR 94, NRPP 95) [27]. Over 3000 research papers show that low-dose irradiation is stimulatory and beneficial in a wide variety of microbes, plants, invertebrates, and vertebrates [34, 35]; this was excepted by France, Japan, and China. The epidemiological studies of irradiated population exhibit reduced risk of cancer from low dose of radiation [36].

Radiation is not only the cause of cancer. Cancer is always a cancer. Lung cancer caused by smoking tobacco is medically identical to lung cancer caused by inhalation of radioactive gas. If a patient suffers from cancer, there is no absolute certainty that the reason was radiation. Even if a cancer patient has received a lifetime dose of 500 mSv which is many times higher than the annual dose limit for professional radiation workers, then it is ten times more likely that his or her cancer was caused by another reason than radiation.

6. Quantities and units used in radiation protection

The physical quantity and radiation protection quantities are used in radiation protection. For the estimation of the effects of radiation and for their control, it is necessary to define a suitable unit for the measurement of radiation dose. Exposure to radiation or radioactive materials results in the irradiation of people. The exposure may be external from the sources outside the body or internal from the sources inside the body. The energy absorbed from ionizing radiation per unit mass of the materials at the point of interest is called "radiation dose."

6.1. Absorbed dose (D)

The amount of energy is deposited in the medium per unit mass. The unit of absorbed dose is gray. The activity is defined as rate of disintegration per second.

In addition to the physical quantity, other units related to the quantities have been introduced to account for biological effects of radiation upon the tissue. The quantities are organ dose, equivalent dose, effective dose, committed dose, and collective dose.

6.2. Organ dose

The mean dose D_T in specified tissue or organ T of the human body is given by

$$D_T = \frac{E_T}{m_T} \tag{10}$$

where E_T is the total energy deposited by radiation to the organ and m_T is the mass of the organ.

6.3. Equivalent dose (H_T)

The ionizing power of different radiations are different; their biological effects on living cells are also different. Biological damage caused by the radiation is a strong function of the specific radiation and its energy. For the same dose to organ, α and neutron will cause more effect compared to γ and β radiation. The difference lies in the linear energy transfer of different particles. To account of this effect, each radiation type is assigned a radiation weighting factor (W_R) or relative biological effectiveness (RBE). The equivalent dose is the product of radiation weighting factor and absorbed dose. The unit equivalent dose is Sievert (Sv):

Equivalent dose (H_T) = average absorbed dose × radiation waiting factor.

$$H_T = \sum W_R D_{T,R} \tag{11}$$

The radiation weighting factor for X-rays and γ and β particle is 1 and for α is 20 and neutron is 5–20.

6.4. Effective dose (E)

The relation between the probabilities of developing the biological effect depends on specific organ or tissue receiving radiation. To accounts for this, a tissue weighting factor W_T, defined for different organs of the body. It is independent on type of radiation and energy. The unit of effective dose is also Sievert. The tissue weighting factors for some organs are given in **Table 6**.

Sl. no.	Tissue or organ	Tissue weighting factor W_T
1	Gonads	0.20
2	Bone marrow	0.12
3	Colon	0.12
4	Lung	0.12
5	Stomach	0.12
6	Bladder	0.05
7	Breast	0.05
8	Liver	0.05
9	Esophagus	0.05
10	Thyroid	0.05
11	Skin	0.01
12	Bone surface	0.01
13	Remainder	0.05
14	Whole body	1

Table 6. Tissue weighting factors [ICRP 1990 recommendations].

$$E = \sum_T W_T H_T \tag{12}$$

6.5. Committed dose

The radionuclide delivers the dose to the tissue or organ in the body which remains in the body called committed dose. The unit is Sievert (Sv)

6.6. Collective dose

This refers to the group of people, who are exposed to radiation from the source and the period of exposure. It is calculated by the product of number of people exposed with average dose from the source. The unit is ManSv.

7. Categories of exposure

ICRP considers three types of exposure [38]:

1. Occupational exposure, which is the exposure incurred at work and principally as a result of work. Exposure to natural sources in the workplace would not normally come under occupational exposure. Exceptions are exposures to high levels of radon in specific workplaces such

as uranium mines, space as identified by the regulatory agency, and exposure of the crew of jet aircraft and space flights to higher levels of radiation from cosmic rays and solar flares.

2. Medical exposure, which refers to exposures incurred by (a) individuals as part of their own diagnosis or treatment, (b) exposures incurred knowingly and willingly by individuals helping in support of such patients, and (c) exposures incurred by volunteers as part of a program of biomedical research.
3. Public exposure, which comprises all other exposures. (The component of public exposure due to natural sources is by far the largest of the total.)

In practices and in intervention, it will often be virtually certain that exposures will occur and their magnitude will be predictable, albeit with some degree of error. However, in "potential exposures," there will be potential for exposure but no certainty that it will occur.

8. Radiation protection, safety, and dose limits

Radiation protection and safety standard sources for occupational and public radiation exposure are established by Basic Safety Standards (BSS) and ICRP recommendations. The system is based on the following general principles:

1. Justification: The practices or a source of radiation exposure is to provide the benefit for the exposed individuals or to the society; otherwise it cannot be considered.
2. Optimization: In relation to any particular source within a practice, the magnitude of individual's doses, the number of people exposed, and the likelihood of incurring potential exposures should be minimum, and dose is as low as reasonably achievable (ALARA). The economic and social factors are being taken into account with the restriction that the doses to the individuals delivered by the source be subject to dose constraints.
3. Dose limitation: The exposure of individuals resulting from the normal and all the relevant practices should be subject to dose limits. In any normal circumstances, the individuals

	Occupational exposure	**Exposure to apprentices 16–18 years of age**	**Public exposure**
Effective dose (whole dose) (mSv)	20, averaged over 5 consecutive years 50 in a single year[a]	6	1, averaged over 5 consecutive years in a single year[b]
Equivalent dose (eye lens) (mSv)	150	50	15
Equivalent dose (hands, feet, skin) (mSv)	500	150	50

[a]Provided that the average effective dose over 5 consecutive years does not exceed 2 mSv/a.
[b]Provided that the average effective dose over 5 consecutive years does not exceed 1 mSv/a.

Table 7. Annual dose limits according to BSS schedule II and ICRP report 60.

should not be exposed to more than the specified dose limits. Not all sources are susceptible to control by action at the source, and it is necessary to specify the sources to be included as relevant before selecting a dose limit.

According to Basic Safety Standards, the dose limit is defined as the value of effective dose or equivalent dose to individuals from controlled practices that shall not be exceeded [39]. The latest recommended limits were specified in the ICRP (60), 1990, and is given in **Table 7**.

9. Control of occupational and public exposure

According the International Labor Office (ILO), the occupational exposure refers to the exposure of a worker that is received or committed during the period of work [40]. Radiation protection of workers is essential for the same and acceptable use of radiation, radioactive materials, and nuclear energy. The IAEA and ICRP frame a norm and regulations to protect the workers. The dose limits for occupational and public exposure are given in **Table 7**.

In medical professionals, to minimize the radiation exposure, one can follow as low as reasonably achievable (ALARA) and personnel shielding options (e.g., two-piece wraparound aprons, thyroid shields, and eye protection) which should be used to effectively attenuate scattered X-ray levels. For medical exposure of patients, dose limit is not appropriate to apply. Therefore medical radiation does not have dose limits and generally used diagnostic reference level (DRL) as a reference value.

To decrease radiation exposure risks, any medical radiation exposure must be justified, and the examinations which use ionizing radiation must be optimized. Justification means that the examination must be medically indicated and useful. Optimization means that the imaging should be performed using doses that are as low as reasonably achievable (ALARA), consistent with the diagnostic task.

The control of public exposure is normally exercised by the application of controls at the source rather than in the environment. According to ICRP recommendations, the dose limits should not exceed 1 $mSvy^{-1}$ (excluding normal background radiation). However, in special circumstances, a higher value can be allowed in a single year, provided that the average over 5 years does not exceed 1 $mSvy^{-1}$.

For controlling the occupational exposure, the following three parameters are considered:

1. Distance: The distance between the source and exposing worker should be large to reduce the amount of radiation received by the workers.

2. Time: The radiation dose is directly proportional to the time spent in the radiation. Therefore the time of exposure should be as small as possible.

3. Shielding: Depending upon the type of radiation, different materials are used for shielding. For gamma radiation high-atomic-numbered elements are used, because the rate of

energy loss is directly proportional to Z^5. For neutron, high absorption cross section and low-atomic-numbered elements are used for shielding; hydrogen and hydrogen-based materials are well suited for neutron shielding. The plastic can be used to form an efficient barrier for dealing with high-energy beta radiation.

10. Conclusions

Radiation is present everywhere, and it is a permanent feature of the environment; thus the risk associated with the radiation can only be restricted and not eliminated entirely. Applications of radioisotopes are growing in industry, agriculture, medicine, and many other fields of industry and research, benefiting humanity. Irradiation is used around the world to preserve food staffs and reduce wastage, and sterilization techniques have been used to eradicate disease-carrying insects and pests. The activities involving radiation exposure are subject to certain standards of safety in order to protect the individuals exposed to radiation, be it occupationally, for medical diagnostic or therapeutic purposes or as the member of the public. Higher radiation dose produces harmful effects; the severity of effect increases with increasing the dose. There is a considerable uncertainty whether the low-dose radiation (LDR) is beneficial or harmful; some scientific groups favor the low dose is harmful, and other scientific groups argued that low dose is beneficial to mankind. At the present low-dose radiation is the thrust area for research.

Acknowledgements

It is a pleasure to acknowledge the help from Rangaswamy D R and Sunilkumar, Research Scholars, Department of Physics, during the preparation of the manuscript.

Conflict of interest

There is no conflict of interest.

Author details

Sannappa Jadiyappa

Address all correspondence to: sannappaj2012@gmail.com

Department of Studies and Research in Physics, Kuvempu University, Shivamogga, Karnataka, India

References

[1] Rejali, et al. Dynamic studies with radioisotopes in medicine. In: Proceedings of an IAEA Symposium; 1970. p. 116

[2] Ann VE, Greg P. Patently female: From AZT to TV dinners: Stories of women inventors and their breakthrough ideas. New York: Wiley; 2002. p. 99

[3] Stassinopoulos EG. The earth's trapped and transient space radiation environment. In: McCormack Percival D, Swenberg Charles E, Bucker Horst, editors. Terrestrial Space Radiation and its Biological Effects. New York: Plenum Press; 1998. pp. 5-35

[4] United Nations Scientific Committee on the Effects of Atomic Radiation (UNSCEAR). Sources Effects and Risks of Ionizing Radiation. Report to the General Assembly, with Annexes. New York, USA: United Nations Publication; 1988. pp. 49-51

[5] International Atomic Energy Agency (IAEA). The Use of Gamma Ray Data to Define the Natural Radiation Environment. Report No. IAEA-TECDOC-566; 1990. ISSN: 1011-4289

[6] Taylor SR. Abundance of chemical elements in the continental crust: A new table. Geochim et Cosmochim Acta. 1964;**28**:1273-1284

[7] Wahl W. Radionuclide Handbook for Laboratory Workers in Spectrometry, Radiation Protection and Medicine. Germany: ISuS; 2007

[8] Mohammed A, El-Hussein A, Ali A. Measurements of Thorium-B (^{212}Pb) in the outdoor environment and evaluation of equivalent dose. Journal of Environmental Radioactivity. 2000;**49**:181-193

[9] Israelsson S, Knudsen E, Ungethum. Simultaneous measurements of radon ^{222}Rn and thoron ^{220}Rn in the atmospheric surface layer. Tellus. 1973;**25**:281-290

[10] Bodansky D. In: Bodansky D, Roskin MA, Stadler DR, editors. Overview of the Indoor Radon and Its Hazard. Seattle and London: University of Washington Press; 1989. pp. 3-15

[11] Browne E, Firestone RB, Shirley VS. Table of Radioactive Isotopes. New York: John Wiley and Sons, Inc; 1986

[12] IAEA. Naturally occurring radioactive material (NORM-V). In: Proceeding of an International Symposium, Seville, Spain. Vol. 47; 2007

[13] Reitz G. Radiation environment in the stratosphere. Radiation Protection Dosimetry. 1993;**48**:5-20

[14] United Nations Scientific Committee on the Effects of Atomic Radiation (UNSCEAR). Sources, Effects and Risks of Ionizing Radiation. Report to the General Assembly, United Nations, New York; 2000

[15] Bartlett DT. Radiation protection aspects of the cosmic radiation exposure of aircraft crew. Radiation Protection Dosimetry. 2004;**4**:349-355

[16] Watson SJ, Jones AL, Oatway WB, Hughes JS. Ionizing Radiation Exposure of the UK Population. Review. Health Protection Agency, Centre for Radiation, Chemical and Environmental Hazards, Radiation Protection Division, Chilton, Didcot, Oxfordshire, OX11 0RQ, UK; 2005

[17] Shahbazi-Gahrouei D. Natural background radiation dosimetry in the highest altitude region of Iran. Journal of Radiation Research. 2003;**44**:285-287

[18] Kwan-Hoong Ng. Non-ionizing radiations-sources, biological effects, emissions and exposure. In: Proceedings of the International Conference on Non-Ionizing Radiation, at UNITEN ICNIR; 2003

[19] Krewski D, Lubin JH, Zielinski JM, Alavanja M, et al. Residential radon and risk of lung cancer: A combined analysis of 7 north American case-control studies. Epidemiology. 2005;**16**:137-145

[20] Abdel Rahman RO. Introduction to Current Trends in Nuclear Material Research and Technology, Ch (1)

[21] Chuvilin Y, Khvostionov VE, Markovskij DV, Pavshouk VA, Zagryadsky VA. Low-waste and proliferation-free production of medical radioisotopes in solution and molten-salt reactors. In: Abdel Rahman RO, editor. Radioactive Waste. Rijeka, Croatia: Intech; 2012

[22] Sahrama AK. Indian Nuclear Society Publications. BARC Mumbai; 2010

[23] Bjorn Wahlstrom. Radiation, Health and Society. International Atomic Energy Agency; 97-05055 IAEA/PI/A56E. Austria, November 1997

[24] Luckey TD, Lawrence KS. Radiation Hormesis; the good, the bad, and ugly. Dose Response. 2006;**4**:169-190

[25] Dawson D, Fleck R, Wadham A, Bird P. WNTD: Radiation Damage to Materials. Jun. 1993

[26] Sandhya Rani N, Sannappa J, Demappa T. Mahadevaiah: Effects of CdCl2 concentration and gamma irradiation on the structural, thermal and electrical conductivity properties of HPMC polymer electrolyte films. IOSR-Journal of Applied Physics. 2014;**6**:30-41

[27] Chapiro A. Radiation Chemistry of Polymeric Systems. New York: John Wiley & Sons; 1962

[28] Nasef MM, Güven O. Radiation-grafted copolymers for separation and purification purposes: Status, challenges and future directions. Progress in Polymer Science. 2012;**37**:1597-1656

[29] Nasef MM, Hegazy ESA. Preparation and applications of ion exchange membranes by radiation-induced graft copolymerization of polar monomers onto non-polar films. Progress in Polymer Science. 2004;**29**:499-561

[30] Chapiro A. Radiation Chemistry of Polymeric Systems. New York: Interscience Publishers; 1962

[31] Spinks JWT, Woods RJ. An Introduction to Radiation Chemistry. Wiley-Interscience; 1990

[32] Schonbacher H, Stolarz-Lzicka A. Compilation of Radiation Damage Test Data. Part I: Cable Insulating Materials. Geneva: CERN; 1979

[33] Singleton, George 1958. Contemporary Authors. Encyclopedia.com. Apr. 2018

[34] Muckerheide J. Low-Level Radiation Health Effects: A Compilation of Data and Programs. Needham, Mass: RSH, Inc.; 2001

[35] Knoll GF. Radiation Detection and Measurement. 3rd ed. Wiley India Pvt. Ltd; 2009

[36] Syed MA, Raziuddin A, Mohammod AK. Application of radioisotopes and radiation in the field of agriculture: Review. Journal of Biological Sciences. 2001;**1**:82-86

[37] Doss M. Linear no threshold model vs. radiation hormesis. Dose Responses. 2013; **11**:480-497

[38] Nagaratnam A. ICRP Recommendations: An Overview, ISRP (K)-BR-6; 1994

[39] International Atomic Energy Agency, International Labour Office, Assessment of Occupational Exposure due to External Sources of Radiation, Safety Standards Series No. RS-G-1.3. IAEA: Vienna; 1999

[40] International Labour Office. Radiation Protection of Workers (Ionizing Radiations), ILO Code of Practice. Geneva: ILO; 1987

Chapter 3

Synthesis of Chemical Elements and Solid Structures in Atomic-Nuclear Reactions in Dense Gas-Metal Systems Irradiated by γ Rays

Roland Wiśniewski

Additional information is available at the end of the chapter

http://dx.doi.org/10.5772/intechopen.78740

Abstract

The new effects observed in dense gas-metals systems (example D_2-Pd) being irradiated by braking gamma quanta with threshold energy a little bellow 10 MeV, for energy about 25 MeV (giant dipole nuclear resonance—GDNR) and effects of irradiation of pure gazes (D_2, He… 10 MeV) is presented in this chapter. The irradiation time in all cases was as large as 10^5 s under large gamma flux intensity conditions (generated by electron current in used accelerators of about 20 μA). Realization of experiments were as follow: chosen gas in room temperature was compressed to high pressure (1–3 kbar) in beryllium bronze pressure chamber or in stainless steel high pressure capillary with chosen metal samples inside of it or without samples. In all cases new elements were detected in relatively large amounts, some of them are light elements, i.e., C, O and the others are heavy elements, i.e., Pb, Bi. At GDNR with Pd, V and other metals, unexpected phenomena as shape changes, "micro-protuberances," micro objects of specific element contents, cracks of specimen surfaces were observed too. The identification of the mechanism that induced these effects is a problematic issue, trans-nuclear molecule and multinuclear reaction concepts could be used to explain these induced effects. The justification of carbon generation in the samples is shortly described.

Keywords: gamma quanta irradiation, dense gas-metal systems, fission, fusion reaction, transmolecule, chemical composition

1. Introduction and historical outline

In the 2005, Didyk [1] from the Join Institute for Nuclear Research in Dubna (RF) was visited the National Centre for Nuclear Research (Otwock-Świerk, Poland). He was introduced to high pressure method for hydrides synthesis [2] by Author, and close cooperation between them was born. In their first experiment they have placed the Pd samples into the gaseous deuterium in increased pressure (3–5 kbar), and exposed this setup to gamma radiation, in order to significantly increase the content of deuterium in palladium (D/Pd > 1) by deuterium atomization and the call to other possible phenomena accompanying referred to the radiation. The electron accelerator Microtron MT-25 in the Flerov Laboratory of Nuclear Reactions JINR in Dubna, was used as the gamma radiation source. Intense braking radiation (using tungsten target, thickness 2.5 mm) with electron current 10–24 μA), long exposure times (E5s) was applied to the setup resulted in the transformation which one of the effects was creation in macro-scale objects (element composition). After this, unexpected results, experiments were continued. The range of threshold energy of gamma-quantum in the greater part of them was around 10 MeV, in a few cases, was increased up to 24 MeV. The latter value corresponds to the scope of the giant nuclear dipole resonance of exposed samples. Initially, the high-pressure chamber material were seen only as the normal processes of absorption of gamma radiation, provided through a special item of apparatus-cork related pane. Researchers conducted on pure gases (no samples in the high-pressure chamber) appeared to be surprising enough to force change of approach to explaining the results. We started to perceive results as the effect of exposure to quanta gamma the selected gases under high pressure, surrounded by metallic shell in the form of a metallic chamber with or without an object located around the centre of shell (see **Figure 1**). This idea was implemented in the form of custom designed high-pressure systems (laboratory), presented in the body of the chapter. The experience described below, you need to request the presence of a number of nuclear processes (fission and fusion) initiated by photo reactions (gamma rays) and certain more or less complicated physical and chemical processes. Effects, in the sense of the masses of generated objects (and their composition) are macroscopic, which allows to talk about the possibility of new nuclear technology (engineering) allowing to produce heavy elements such, as rare earth elements. Work related to the construction of high-pressure apparatus, developing of different measurements systems, gas filling under high pressure, preparation of samples, formal documentations took place at Pressure Investigation Laboratory of the National Centre for Nuclear Research in Poland. The high-pressure equipment was built by Author and PhD. Wilczyńska-Kitowska T. Irradiation procedures were carried out in the Joint Institute for Nuclear Research in Dubna, in the Flerov

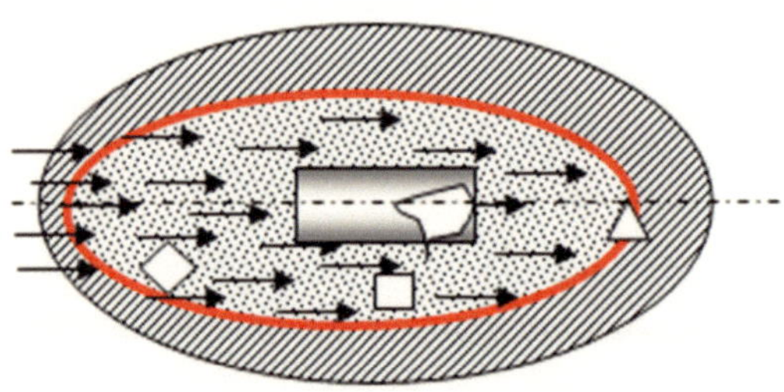

Figure 1. Simplified model of conducted experiments. The arrows presents high energy gamma quanta stream passing through the inlet port of the pressure shell. Polygon character symbolizes the central object changes. Square character symbolizes objects which were observed inside of the chamber and triangular character changes in inner element (red line) separates the gas from the pressure shell.

Laboratory of Nuclear Reactions. Investigation of obtained objects, using scanning electron microscopy (SEM) and X-ray micro sound analyzer (XRMPA) were performed in the Analytical Centre of the Skobeltsyn Laboratory of Lomonosov Moscow State University and at the Research Institute for Perspective Materials and Technologies in Moscow. We have observed: macroscopic surface changes, the emergence of new elements in the macroscopic amount, change the shape of the samples, micro eruption from the metallic surface and exposed to weight changes of investigated samples.

In order to validate our results a number of additional experiments were conducted: (A) the main investigation element—palladium sample—was subjected to operation of pure deuterium gas at a pressure of approximately 20 kbar over a period of several months. After dismounting, no changes resembling result received after irradiation of the similar setup, in the object itself or chamber were found. The currently known classic effects (volume change, resistance, etc.) associated with the process of saturation Palladium by deuterium gas under pressure was tested by the author much earlier [3]. (B) Irradiation of the sample of palladium, without presence of high-pressure gas did not resulted in any changes on the surface of the sample (typical radiation effects such as structural defects, were not taken into account). (C) High-pressure chamber, with the sample of palladium, and average vacuum after irradiation did not show any of our research typical effects. (D) At lowered deuterium pressure up to 60 bar, irradiation time 1.3E5s, electron current 1.1E14 s^{-1} with typical Pd specimen (purity of 99.997%) no effect was observed either.

In authors opinion, farther studies of presented results and proposed interpretation, should result in new specific methodology of basic investigation, and new technology in the future. Since, known classic nuclear effects and theories (and possible physical-chemical processes) are fundamental for interpretation of observed phenomena by author and his co-operators but others new ideas are desirable. Why so relatively small pressure (1000 bar) of gases play so essential role? Does this "trigger" pressure value differs for different gases?

Simplified model of conducted experiments is shown in **Figure 1**.

The arrows presents high energy gamma stream passing through the inlet port of the pressure shell. Polygon character symbolizes the central object changes. Square character symbolizes objects which were observed inside of the chamber and triangular character changes in inner element (red line) separates the gas from the pressure shell. Similar objects can be exist in similar condition in some time in somewhere is Cosmos. What effects would took place there we can see as results of our experiments described below.

The following experimental investigations were carried out:

1. Simple system Pd-D_2, classic type of high-pressure chamber (CTHPC) [4].

2. Simple system Pd-D_2, (CTHPC), Pd-D [5].

3. Simple system Pd-H_2, (CTHPC) [6].

4. Double system Pd-Re-D_2, finger type HPC, giant dipole nuclear resonance (GDNR) [7].

5. Double system, Al-YMn2-D_2, FTHPC, GDNR [8].

6. Double system, V-SS-D_2, CTHPC, GDNR [9].

7. Complex system, Al-YMn2-Al-YMn2-Cu-SS-D2, FTHPC, GDNR [10].

8. Complex system, Sn-Mo-Fe-Ni-Bi-Ta-Cu-H_2, CTHPC [11].

9. Pure Helium, CTHPC [12–14].

10. Pure Hydrogen, CTHPC [15].

11. Pure Deuterium, CTHPC [16].

In this chapter, as examples of gamma irradiation investigations in metal-dense gases systems: Pd-D_2 (10 MeV), Pd-Re-D_2 (24 MeV) and pure H_2, D_2 and He (10 MeV) are presented.

2. Experimental setup

A hydrogen (deuterium) high-pressure chamber (H(D)HPC) has been designed. The first high-pressure apparatus for gaseous hydrogen (deuterium) was constructed and build by Author [3]. Presented here, two types, apparatus are portable and adapted to γ irradiation. The scheme of the DHPC with deuterium and Pd specimen inside, used by us in first experiments, is presented in **Figure 2**. Inside the DHPC with an inner cylinder of diameter of 0.4 cm there was placed a cylindrical Pd-rod (with 99.98% purity) of diameter 0.38 cm and length 0.5 cm. A distance manganin sleeve of length 0.6 cm was disposed between the Pd-rod and brass screw of length 0.5 cm and diameter 0.4 cm. The DHPC inside volume was equal to $V_D = 0.264\ cm^3$ and filled with molecular deuterium under the pressure of about 3 kbar. At such pressure the molecular deuterium density was about $n_{D_2} = 2.593 \times 10^{22}$ mol.D_2/cm^3 (see [26]). Therefore, the overall number of deuterium atoms in the DHPC inner volume was about ND = 1.4×10^{22} at. D. In **Figure 3** a last form of this apparatus is presented. Main changes concern to elimination of any plastic type O-ring system and create full CuBe brass environment of investigated metal-gas system.

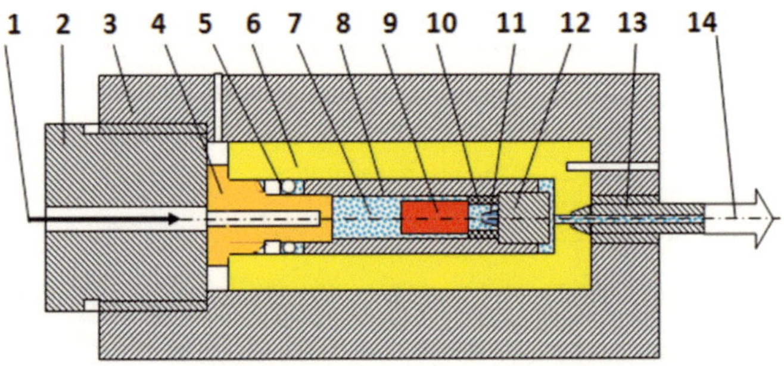

Figure 2. The schematic drawing of high-pressure apparatus (DHPC). 1—γ-quanta, 2—closing screw with hole, 3—reinforcing high-pressure chamber body, 4—$Cu_{0.98}Be_{0.02}$ "window" plug, 5—high-pressure seals, 6—$CuBe_2$ high-pressure chamber, 7—(hydrogen) deuterium under high pressure, 8—brass sleeve, 9—investigated Pd-rod, 10—distance manganin sleeve, 11—expected reaction product, 12—brass screw, 13—high-pressure-connecting capillary, 14—high-pressure valve, strain gauge pressure sensor and gas filling inlet [4].

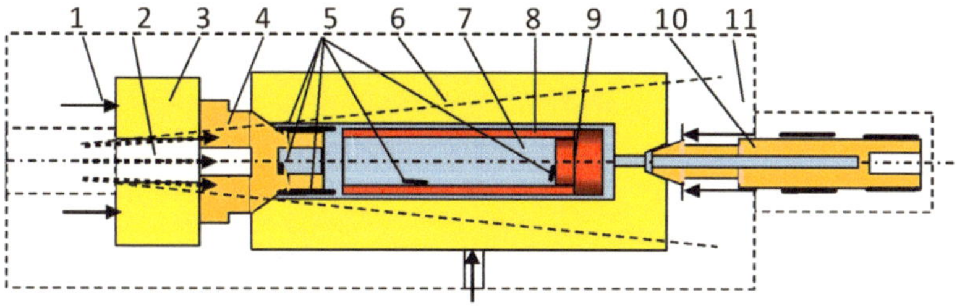

Figure 3. Elements of modified high-pressure apparatus exposed to irradiation by gamma rays: 1—Pressure for sealing of the closing screw to the window plug, 2—Gamma ray flux of initial diameter of 5 mm, 3—$CuBe_2$ element inserted in the closing screw, 4—Gamma ray window-plug, 5—surface regions where carbon foils and other particles were observed in large quantities, 6—high-pressure chamber reinforced by outer element, 7—gas (under pressure), 8—pure copper (or $CuBe_2$) sleeve, 9—pure copper (or $CuBe_2$) closing element, 10—gas supplying and pressure measurement strain gauge device, 11—schematically shown reinforcing outer element.

Figure 4 shows part of this DHPC adapted for irradiation by γ-quanta of sets of specimens from various materials in the shape of wires with relatively small diameters. For this apparatus the walls were made of 204 stainless steel (204SS) to exclude such chemical elements, observed in the first experiment, as copper and zinc. These elements comprised the degraded layer of thickness over 80 μm in palladium as in the first experiment [4].

Two type experiments to study γ-quanta irradiation effects were carried out at the electron accelerator MT-25, FLNR, JINR using two electron beams of energies 9.3 and 25 MeV, which correspond to γ-quanta energy < 8.8 and 23 MeV, respectively. The electron current at the tungsten convector shaped as a disk of 40 mm diameter and 2.5 mm thickness used to transform the electron flux into gamma rays was (1–1.5) × 10^{14} electron/s. It can be supposed that generated gamma flux is the same of order. The electron beam was 6–7 mm in diameter. The beam spread of gamma quanta at the intensity half-height amounted to 10 ± 1° in horizontal and 8 ± 1° in vertical direction. Right behind the convector, an electron absorber of 25 mm thickness made from D16T duralumin was placed.

The γ quanta spectra Nγ(Eγ) were calculated using the expressions presented in [17].

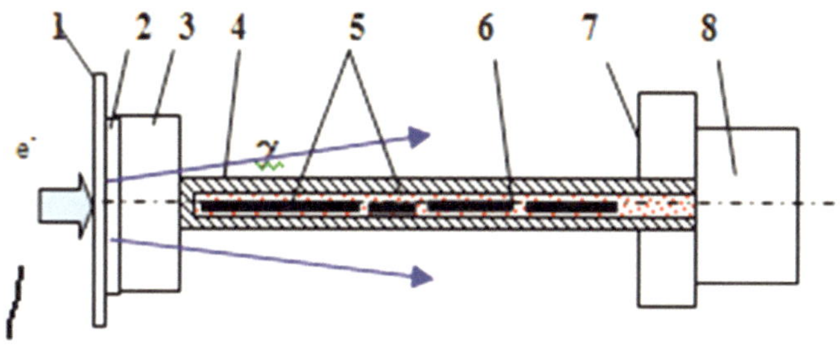

Figure 4. A scheme of capillary type DHPC intended for irradiation of materials: 1—vacuum system wall, 2—tungsten target, 3—alumina absorber, 4—capillary type pressure chamber, 5—specimens: One Pd and three of different length Re, 6—gaseous deuterium, 7—safety wall, 8—gas fill up, pressure and temperature measurement systems.

Calculation of the energy spectrum of γ-quanta from the units of measurement in MeV^{-1}·$steradian^{-1}$ into the flux density of γ-quanta in cm^{-2} and per area of the input window plug at the DHPC-target gave value about 3%. We can now calculate [18], for energy intervals $\Delta E_\gamma = 1$ MeV, the number of neutrons and protons produced by γ-quanta in the photofission reaction D(γ,n)p—mostly expected in case of irradiated D_2—with the maximal energies $E_\gamma^{max} = 8.8$ MeV and $E_\gamma^{max} \leq 23.0$ MeV using the expression:

$$Y(E_\gamma^{max}) = \beta \cdot N_D \cdot \int_{\Delta W}^{E_\gamma^{max}} \sigma_{D(\gamma,n)p}(E_\gamma) \cdot N_\gamma(E_\gamma) dE_\gamma \quad (1)$$

where ΔW = 2.22 MeV is the deuteron bond energy. In [4] adequate cross sections, histograms with neutron and proton yields $N_n(E_\gamma)$ and $N_p(E_\gamma)$, respectively, at two ranges $n_H \approx 8.8$ Mev and $p_H \approx 23$ Mev are presented for two energies of electron beams of 1 μA.

Developed measurement of pressure technique (pressure sensor, with strain gauge elements shown in **Figure 3**) was used in our condition continuously and with sufficient high accuracy. Temperature measurements were realized using standard thermoelement equipment. Irradiations processes performed in our experiments were fully controlled by Laboratory Security Service as normal procedure in such type experiments.

3. Induced effects in gas (D2) and metal-gas systems (Pd-D2)

Here is presented information based on the first experiment [4], i.e., on irradiation process of Pd-D_2 system under pressure of gaseous deuterium about 3 kbar. During the whole time of DHPC exposure to γ-quanta $t \approx 2.22 \times 10^4$ s, the average current of the electron beam was ≈ 7 μA. The total number of neutrons and protons produced by γ-quanta per experiment $N_{n,p} = 3.23 \times 10^{10}$ n, p. During the irradiation by γ-quanta the DHPC was cooled by a flux of compressed air at the temperature ≈ 20°C, nevertheless, the temperature of the DHPC external surface significantly exceeded 100°C.

Prior to opening of the DHPC, some hold-up time aimed to decrease induced activity was allowed to elapse before the chamber was finally opened. The Pd-rod with the external initial diameter 3.8 mm was found to be wedged due to its increased overall dimensions inside the brass sleeve with the inside diameter d = 4 mm. After 3 days the palladium specimen decreased in size and was easily extracted. Relative diameter change ((0.2/4)100% = 5%) was caused by other than the normal volume change related to beta phase appearance in PdD system.

3.1. Synthesized novel object (SNO)

On the flat inside surface of the brass screw (see **Figure 2**, p. 12), there was discovered a synthesized novel object (SNO). The optical (a) and SEM (b) scanning electron microscopy images of this object are presented in **Figure 5**. Because of high value of dielectric permeability, the surfaces of the SNO object and brass screw were covered with a thin gold layer of thickness up to 1000 Å (see **Figure 5b**).

http://dx.doi.org/10.5772/intechopen.78740

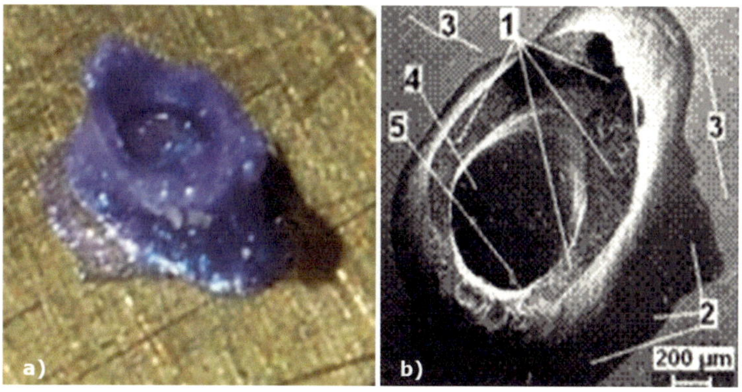

Figure 5. A view photo of the brass screw with the synthesized novel object (SNO, informal name Annavit [12]) (a) and "volcano" image with the SNO made by scanning electron microscopy (b) [4].

Let us present concentrations of chemical elements (at.%) obtained by microelement analysis from the surface brass around the SNO object ($3_{1,2,3,4,}$ **Figure 5b**). They are as follows (characteristic data): C^6 - 39.70, O^8 - 0.86, Fe^{26} - 0.25, Cu^{29} - 29.29, Zn^{30} - 18.40, Nb^{41} - 1.49. Concentrations of chemical elements (at.%) gained by microelement analysis from the surface in the spread melt "lava" near the SNO ($2_{1,2,3,}$ **Figure 5b**) are as follows: (characteristic data): C^6 - 55.01, 0^8 - 35.08, Mg^{12} - 1.48, Al^{13} - 0.92, Si - 2.26, K^{19} - 0.35, Ti^{22} - 2.47, Cu^{29} - 0.85, Zn^{30} - 0.49, Au^{79} - 1.10.

The element composition (wt.%) on the top of SNO object for four measured points ($1_{,1234}$ **Figure 5b**) is as follows:

1.1: C^6 - 26.81, 0^8 - 37.78, Mg^{12} - 0.44, Al^{13} - 5.27, Si^{14} - 6.88, K^{19} - 1.7, Ca^{20} - 0.31, Ti^{22} - *13.17*, Fe^{26} - 0.31, Cu^{29} - 1.17, Zn^{30} - 0.67, Au^{79} 5.49;

1.2: C^6 - 50.5, 0^8 - 38.43, Mg^{12} - 1.62, Si^{14} - 2.92, Ti^{22} - 0.32, Zn^{30} - 0.81, Au^{79} - 5.39;

1.3: C^6 - 38.51, 0^8 - 26.35, Mg^{12} - 0.74, Al^{13} - 4.81, Si^{14} - 7.36, Cl^{17} - 0.27, K^{19} - 1.76, Ti^{22} *9.73*, Cu^{29} - 0.7%, Zn^{30} - 0.83, Au^{79} - 8.93;

1.4: C^6 - 56.69, 0^8 - 30.39, Mg^{12} - 1.7, Al^{13} - 0.89, Si^{14} - 4.03, K^{19} - 0.28, Ti^{22} - 2.54, Zn^{30} - 0.67, Au^{79} - 2.81.

As one can see, at two measured points the high concentration of Ti (13.17 and 9.73 wt.%) is accompanied by inclusion of associated elements as Al (5.27 and 4.81 wt.%), Si (6.88 and 7.36 wt.%) and K (1.7 and 1.76 wt.%), and also such elements as Fe, Ca, Cl and Mg.

All the elements observed on the flat top of the "volcano with a crater at the center" object were present on the molten patch or the so-called "lava," although in changed quantities are as follows: Ti—4.44 wt.%, Al—1.04 wt.%, Si—2.65 wt.%, Mg—1.77 wt.% and higher concentrations of Cu and Zn which form the basis of the brass screw, i.e., Cu—2.26 wt.%, Zn—1.32 wt.%. This can be explained by that the spread layer is thin and the electron beam partially reaches the brass substrate during the analysis.

Some various objects such as re-crystallized and melted pieces of brass, as flat thin plates with various forms and small pieces of Pd particles were observed at the bottom of crater (point 4,

Figure 5b). It is necessary to note that all these objects are weakly bond with the floor of crater again with flat object.

The element composition (at.%) of one planar plate-like objects (p. 5 **Figure 5b**), located on the "crater" floor, is as follows: C^6 - 19.57, 0^8 - 20.29, K^{19} - 3.98, Ti^{22} - 50.65 at.% (main elements). The "crater" floor itself has the same chemical composition as these isolated, floor-covering plate-like objects.

On the crater floor one observes the smaller sphere-like objects also complex objects and multiple plate-like objects of size 5–10 μm. The plate-like objects comprise large quantities of Ti—66.42 ± 0.97 wt.% contained as rutile (TiO_2), as was shown by X-ray analysis. It should be noted that on the measured patches of the crater top there was observed a substantial amount of K—4.26 ± 0.16 wt.%, which is close in nucleus charge to titanium, and Fe—1.17 ± 0.07 wt.%, as well as small quantities of Cu—3.68 ± 0.12 wt.% and Zn—2.54 ± 0.10 wt.%. The copper and zinc concentrations in the plate-like objects can be compared to their concentrations on top of the SNO crater walls and in the alloy on the brass screw surface.

The element composition (at%) of the analyzed surface of another complex recrystallized object form on the floor of crater (p. 4, **Figure 5b**) is as follows: C^6 - 20.87, 0^8 - 8.80, Al^{13} - 3.31, Ti^{22} - 2.76, Fe^{26} - 1.35, Cu^{29} - 35.56, Zn^{30} - 21.36, Au^{79} - 2.85.

Then the more detail repeated measurements of brass surface with SNO (**Figure 2** p. 11) were made along the brass screw (**Figure 1** p. 12) diameter for check of obtained results using another electron microscope with X-ray analysis.

Screw surface containing congealed "lava from the crater" together with the chemical composition measured at six points from the right edge of the brass screw towards the melt were measured too [4].

3.2. Surface objects and element composition of Pd specimen

The surface studies of both Pd rod edges and side were carried out before (left positions) and after irradiation (right positions) is presented in **Figure 6**. The deformed Pd sample and the fantastic surface object of the irradiated Pd-rod edge which was close to SNO was obseved.

The element analysis (wt.%) of the Pd-rod edge with seven measured points (**Figure 7**, left and right spectra) is as follows:

Figure 6. Pd-rod specimen before and after irradiation (one can see it layered structure) and black surface view, which was close to SNO.

Figure 7. SEM image of the irradiated by γ-quanta edge of the Pd-rod close to SNO [4].

Ll: 0^8 - 3.05, Cu^{29} – 49.25, Zn^{30} - 37.99, Pd^{46} - 1.51,

L2: Cu^{29} - 12.55, Zn^{30} - 5.92, Pd^{46} - 74.26,

L3: 0^8 - 14.64, Si^{14} - 0.65, S^{16} - 1.08, Ca^{20} - 0.71, Ti^{22} - 1.38, Cu^{29} - 8.29, Zn^{30} - 3.81, Pd^{46} - 25.66,

R1: Cu^{29} - 6.45, Zn^{30} - 3.02, Pd^{46} - 88.18,

R2: Cu^{29} - 3.88, Zn^{30} - 2.55, Pd^{46} - 93.58,

R3: 0^8 - 22.76, Al^{13} - 3.37, Si^{14} - 3.4, K^{19} - 0.54, Ti^{22} - 11.61, Cu^{29} - 31.20, Zn^{30} - 10.41, Pd^{46} - 10.71,

R4: 0^8 - 2.89, Mg^{12} - 0.38, Cu^{29} - 28.73, Zn^{30} - 7.59, Pd^{46} - 52.49.

The entire Pd-rod object turned into an inhomogeneous one, composed of separate clusters with different element compositions. That is precisely why it was impossible to remove the Pd specimen right upon opening the DHPC chamber when this specimen had significantly grown in volume. In the above case, the palladium temperature was such that it enabled processes which brought about formation in some places of copper and zinc concentrations of up to 31.20 and 10.41 wt.% (R3), 49.25 and 37.99 wt.% (L1). At the same time, the concentration of Ti was 1.38 wt.% (L3) and in some places proved to be up to 11.61 wt.% (R3) while for the associated elements it was as follows: Si—0.65 wt.%, S—1.08 wt.%, Ca—0.71 wt.% (L3), as well as for other observed elements: Al—3.37 wt.%, Si—3.4 wt.%, K—0.54 wt.% (R3).

4. Induced effects in binary metal-gas system (Pd-Re-D2)

The goal of this part of chapter is to present the peculiarities of nuclear reaction processes in two or more pure metals placed together in dense gaseous deuterium, saturated with deuterium at high pressures and then being under the action of γ-quanta with the boundary energy 23 MeV, namely, in the region of the giant nuclear dipole resonance.

Palladium specimen (99.96%, in the shape of a rod of diameter 1 mm and length 111 mm) and three rhenium specimens (99.97%, three rods of diameter 1 mm and length 13, 25 and 33 mm)

were placed in the capillary type DHPC under the pressure of gaseous deuterium of about 3.0 kbar. In this experiment irradiation by γ-quanta with the threshold energy 23 MeV was carried out during 19.5 h ($t = 7 \times 10^4$ s) at the average electron beam current (11–12)μA, $(0.5–0.7) \times 10^{14}\ s^{-1}$.

Calculation of γ-quanta spectra and fluxes as well as neutron yield from the d(γ,n)p neutron photodisintegration reaction is described in detail in [4]. Intensity of gamma quanta in different points of investigated specimens is not easy to definite because of its mass absorption and space relation. For sure intensities in introduction surface of specimen and end one may differ several times.

Before opening the DHPC (**Figure 4**), the chamber pressure was measured to be 2860 bar, so definite jump down of pressure was noted. The palladium wire and three rhenium specimens placed in the chamber could be extracted only 3 days later, apparently, after deuterium desorption. The specimens, like the DHPC itself, proved to be highly activated, which did not permit X-ray microelement analysis to be performed even after a long period allowed with the aim of decreasing the induced activity: 4 months for the palladium specimens and over 8 months for the rhenium ones.

4.1. Experimental results for the palladium rod

In **Figure 8** SEM images show places of entry (a) and exit (b) of γ-quanta for the palladium rod. As it is seen, the right edge of the specimen has undergone particularly considerable changes, namely, formation of an elongated part tapering towards the end from 1 mm to 0.461 mm in diameter with a strongly changed (eroded) frontal edge about 1000 μm wide. Both at the left and right ends of the wire one can observe smaller-scale inhomogeneous in the shape of congealed blowouts from the surfaces.

Figure 9 shows congealed-solar-protuberances structures: "a"—blowout from the left surface in the direction opposite to the γ-quanta flux direction, "b"—blowout in the direction coinciding with the γ-quanta flux direction, and "c"—"pedestal" crystal structure on the surface from the direction of γ-quanta penetration. In **Table 1** there is shown element concentration at points 1–3 of observed elements on **Figure 9**. **Figure 10** provides two other interesting photos of blowouts from the end surface from the direction coinciding with the γ-quanta flux direction. In **Table 1** there is shown element concentration at points 4–5 of observed elements on **Figure 10**.

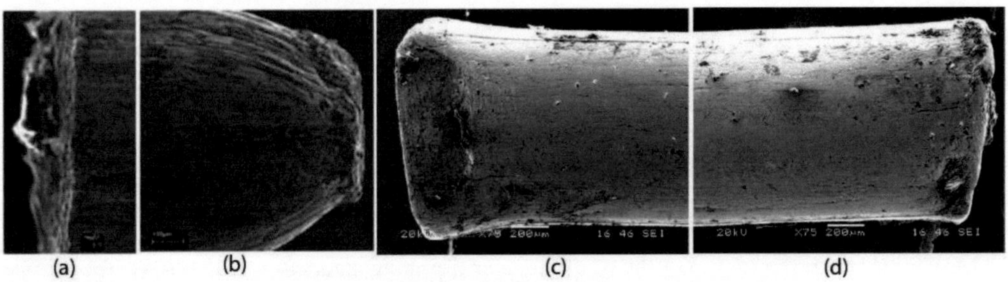

Figure 8. SEM images of the left and right ends of the Pd-rod and two sides of a Re-rod: left—place of entry of γ-quanta, right—in reverse side both specimens after irradiation at GNDR condition in D_2 gas.

http://dx.doi.org/10.5772/intechopen.78740

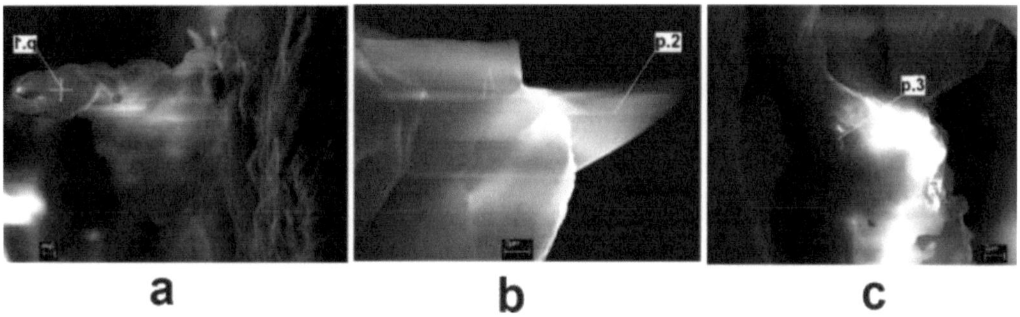

Figure 9. Congealed blowouts: from the γ-quanta entry surface (a), from second frontal surface of the palladium rod (b) and the "pedestal" of structure (c).

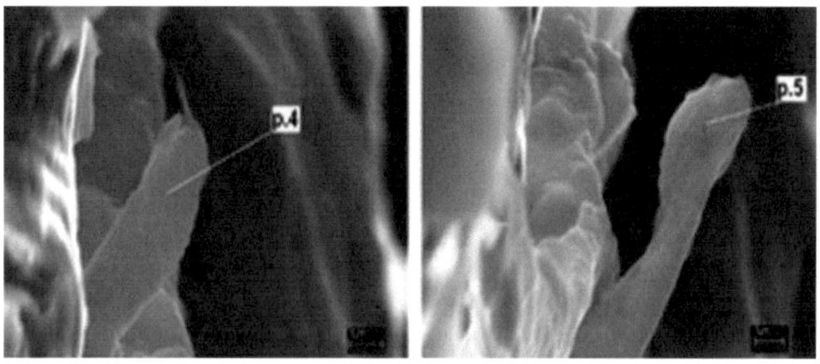

Figure 10. Two congealed blowouts on the direction of γ-quanta escape, from the right end of the palladium rod.

As is shown, the element composition of the blowouts is comprised of a series of elements which are lighter than palladium.

Figure 11 shows SEM images of two surface patches, also near the tapering part (see **Figure 8b**), with several particles on the palladium surface analyzed using XMA (**Table 2**).

Figure 12 shows SEM images of the crack and small crystals on the side surface of the palladium wire. **Table 3** shows element concentration of microcrystal and crack.

4.2. Experimental results for a rhenium rod

It should be noted first and foremost that it was possible to carry out SEM and XMA studies only by using a rhenium wire 13 mm long because it had been placed at a distance of 110 mm (behind the palladium wire) from the back wall of the DHPC entrance window, had a smaller size and, consequently, had acquired a lower induced activity. The other rhenium specimens (of the length 25 and 33 mm) were not accessible for study.

Figure 8 provides SEM images of a rhenium wire with the place of entry of γ-quanta (c) and its reverse side (d), respectively. We can compare the large deformations of both specimens Pd and Re.

Element	Point 1	Point 2	Point 3	Point 4	Point 5
	at.%	at.%	at.%	at.%	at.%
C	—	76.22	66.42	47.58	—
O	27.95	15.77	14.76	11.42	18.19
F	—	—	6.28	—	1.59
Na	—	0.53	1.16	1.02	1.85
Mg	2.68	0.26	—	38.42	58.55
Al	54.56	6.26	7.95	0.45	1.51
Si	3.28	0.07	—	0.07	0.25
S	0.95	0.05	—	0.39	0.58
Cl	0.41	0.19	0.11	0.20	0.44
K	1.42	0.16	1.46	0.13	3.86
Ca	0.52	0.03	0.24	0.10	0.43
Cr	0.54	0.08	0.15	0.12	028
Mn	0.33	0.02	—	0.28	1.06
Fe	2.37	0.34	0.75	0.01	0.44
Ni	—	0.02	—	—	0.97
Cu	1.87	0.01	—		
Zn	1.03	—	—		
Pd	0.23	—	—		
Re	1.86	—	0.74		

Table 1. Element and its concentration in blowouts shown in **Figures 9** and **10** at points 1–5.

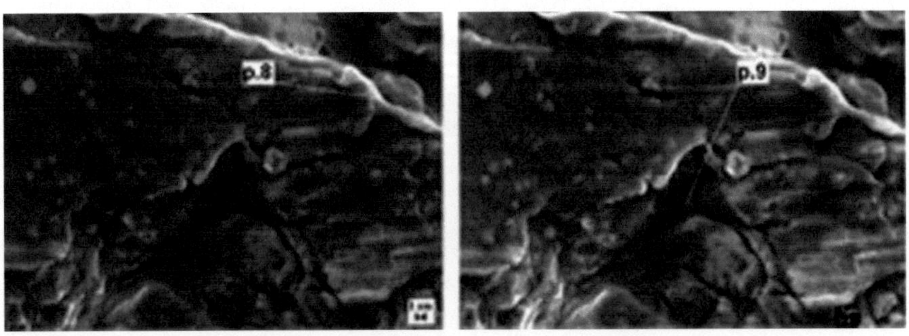

Figure 11. SEM images of a surface patch near the tapering part from **Figure 8b**.

Figure 13a shows a SEM image of surface patch of a Re crack. Once again it is of interest to estimate the concentrations of impurities produced in nuclear reactions from the most degraded cavities, i.e., from the cracks one of which is presented in **Figure 13b**. Averaged XMA analysis was carried out over three surface patches. Its results are summarized in **Table 4** (Spectra 1–4 mean points 1–4). As one can see, the entire rhenium surface is coated with a

Element	Point 8	Point 9
	at.%	at.%
C	37.3	43.8
O	-	27.5
F	-	5.4
Na	-	1.0
Al	1.1	0.8
Si	-	0.4
P	-	0.3
Ca	-	0.7
Cr	0.5	-
Mn	-	0.1
Fe	0.8	0.2
Pd	60.3	17.1

Table 2. Element and its concentration in blowouts shown in **Figure 11** at points 8 and 9.

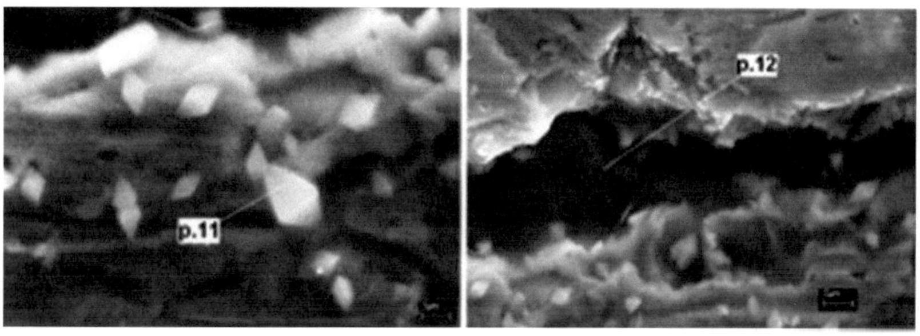

Figure 12. SEM images of the crack and small crystals on the side surface of the palladium wire.

Element	Point 11	Point 12
	at.%	at.%
O	72.1	55.1
Na	1.0	-
Al	0.6	0.7
Si	-	26.9
S	0.2	1.0
Cl	0.1	0.9
K	5.8	1.3
Ca	-	1.8
Mn	0.6	0.8
Fe	0.8	2.3
Ni	-	0.3
Zn	-	0.3
Pd	12.2	8.4
Re	6.6	-
Pt	-	0.2

Table 3. Elements and it concentration crystal deposed in crack vicinity on side surface of palladium specimen (see **Figure 12**).

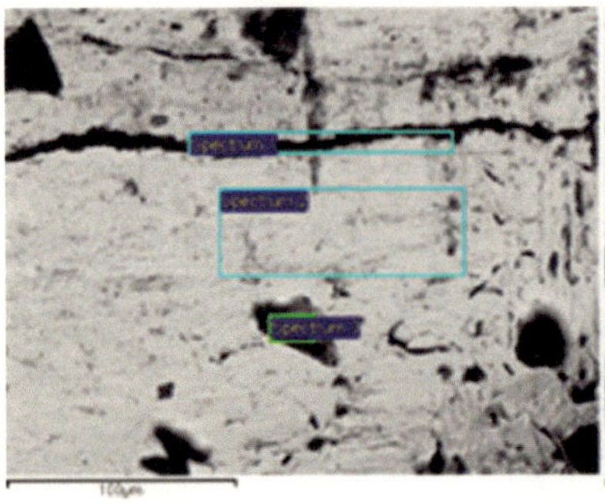
a (spectra 1-3)

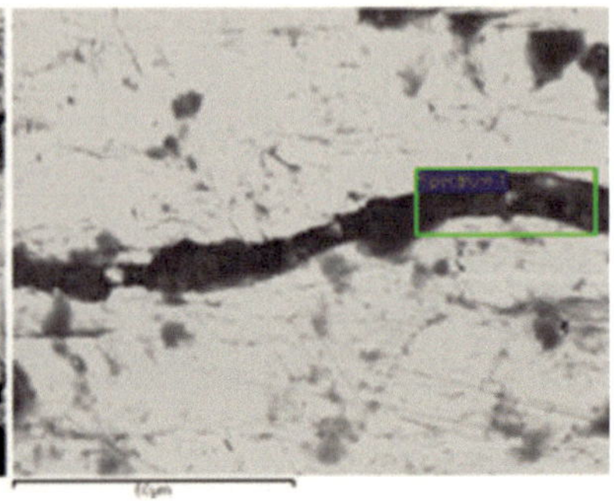
b (spectrum 4)

Figure 13. The surface structure of a Re-rod with marked patches for XMA analysis.

Element	C	O	Mg	S	K
Initial	–	–		–	–
Point 1	65.65	17.44		–	–
Point 2	80.94	13.35		0.25	0.10
Point 3	72.95	14.37		–	–
Point 4	72.22	18.90	0.87	0.39	–
Element	Ca	Fe	Re	Os	
Initial	–	–	100	–	
Point 1	–	–	16.07	0.84	
Point 2	–	0.16	4.92	0.28	
Point 3	–	–	11.91	0.77	
Point 4	1.15	–	6.48	–	

Table 4. Elements and it concentration in Re specimen in places shown in **Figure 13**.

carbon layer, with the carbon concentration reaching 80.94 at.%, a rhenium concentration of only 4.92 at.% and relatively high concentration of osmium up to 0.84 at.%.

5. Induced effects in pure gases

Because of some doubts concerning to our results with deuterium-Me systems, by some typical nuclear physicists, analogous investigations aimed at studying the possibilities of nuclear reactions were carried out using hydrogen high pressure chambers (HHPC) in presence of palladium [6] and tin rods, as well as with hydrogen only without any metallic samples inside [15], under irradiation with 10 MeV braking γ-rays. The phenomenological-model approaches for the description of the obtained experimental data on the basis of nuclear fission reactions

http://dx.doi.org/10.5772/intechopen.78740

within the liquid-drop nuclear model and nuclear fission and fusion reactions are presented in Refs. [18, 19].

The goal of this part of chapter is to present results on the synthesis of chemical elements under irradiation of dense molecular H_2, D_2 and atomic He gases, in the absence of metallic samples, with γ-rays of energies near 10 MeV, which has led to the formation of considerably large particles due to physical chemical reactions.

5.1. Results of γ-ray irradiation of the HHPC filled with molecular hydrogen under 1 kbar pressure

Such pressure corresponds to the atomic concentration of hydrogen $n_D \approx 2952 \times 10^{22}$ at.H·cm^{-3} at its density $p_D \approx 0.0494$ g · cm^{-3} (see monograph [20]). Only a manganin foil was placed in the HHPC chamber, and there was no Pd-rod inside. The chamber irradiation was carried out during 14 h (5.04×10^4 s). The electron beam energy was equal to 9.7 MeV at the average current (20–21)μA. During opening of the HHPC, it spilled out eight small particles about 1 mm in size. Some of them were dark, and the other was of light color. When being photographed with a special tool, all of the particles turned out to be transparent.

The photos in **Figure 14** obtained using a special photomicrography tool show the "light" (a) and "dark" (b) particles at about the same scale.

Figure 15 shows SEM images of a dark particle with a size about 703 μm to 628 μm (a) and of its lengthy "tail" (b). Afterwards, this tail was broken during the studies. In regions 17 (a) and 18 (b) from **Figure 15a** and **b**, X-ray microprobe spectrometer (XRMPS) was used.

Table 5 provides the element compositions measured in regions 17 (**Figure 15a**) and 18 for a dark particle from **Figure 15b**. Note that both parts of the dark particle contain only light elements, such as $_6$C, $_7$N↑, $_8$O, $_9$F↑, $_{10}$Ne↑, $_{11}$Na, $_{12}$Mg, $_{13}$Al, $_{14}$Si, $_{15}$P, $_{16}$S, $_{17}$Cl, $_{18}$Ar↑, $_{19}$K, $_{20}$Ca, and a heavier metal $_{29}$Cu. Here is a series of elements with a nucleus charge change by $\Delta Z = 1$ from $Z = 6$ (carbon) to $Z = 20$ (calcium), where the volatile elements $_7$N↑, $_9$F↑, $_{10}$Ne↑, $_{18}$Ar↑, which could not have formed any chemical compounds, were not detected. The same refers to the elements which are lighter than carbon and have a nucleus charge $Z < 6$, i.e., such elements as $_1$D, $_2$He, $_3$Li, $_4$Be and $_5$B, since they cannot be registered by an X-ray microprobe spectrometer.

Figure 14. Photos of the "light" (a) and "dark" (b) particles obtained in H_2 gas experiment.

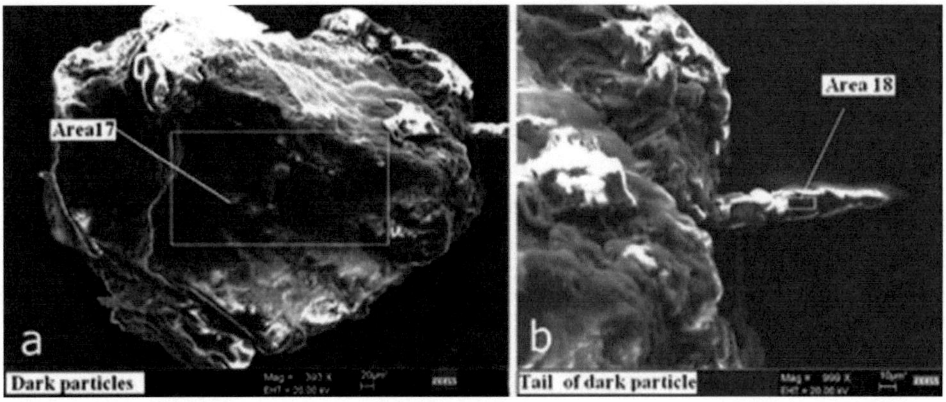

Figure 15. SEM images with a general view of a dark micro particle (a) and its right part with a "tail" (b).

	Figure 15a		Figure 15b		Figure 16a		Figure 18b		Figure 19a		Figure 19b		Figure 21, p. 3		Figure 21b	
El.	wt.%	at.%	wt.%	at.%	wt.%	at.%	wt.%	at.%	wt.%	at.%	wt.%	at.%	wt.%	at.%	wt.%	at.%
C	13.0	17.40	25	32.02	47	64.48	43.3	69.30	11.2	17.15	47	66.95	34.2	46.54	25.6	42.44
N													8.1	9.45	5.16	7.34
O	76.7	77.35	68	65.09	26.6	27.31	12.9	15.47	58	66.42	23.1	24.66	37	38.13	26.1	32.65
F							—	—					—	—	—	—
Na	2.49	1.75	1.05	0.70	—	—	—	—	0.62	0.49	—	—	0.23	0.16	—	—
Mg	—	—	0.03	0.02	1.76	1.19	—	—	6.21	4.68	—	—	0.06	0.04	0.51	0.25
Al	0.19	0.11	0.17	0.10	0.24	0.15	0.03	0.02	0.27	0.18	0.11	0.07	0.07	0.04	2.87	2.25
Si	0.22	0.13	0.11	0.06	1.42	0.83	1.03	0.71	1.27	0.83	0.05	0.03	0.10	0.06	—	—
P	0.16	0.08	0.06	0.03	—	—	0.39	0.24	0.02	0.01	0.04	0.02	0.23	0.12	0.30	0.20
S	—	—	0.03	0.02	0.14	0.07	—	—	0.16	0.09	0.12	0.06	0.28	0.14	0.48	0.30
Cl	4.12	1.88	0.45	0.20	0.18	0.08	—	—	0.50	0.26	0 .07	0.03	0.91	0.42	0.89	0.50
K	1.88	0.78	1.37	0.54	0.23	0.09	—	—	0.37	0.17	—	—	0.80	0.33	0.89	0.46
Ca	1.29	0.52	2.18	0.84	0.21	0.09	1.44	0.69	21.0	9.60	1.91	0.82	0.17	0.07	0.87	0.43
Cr	—	—	—	—	—	—	0.31	0.12	—	—	—	—	—	—	—	—
Mn	—	—	—	—	—	—	0.35	0.12	—	—	—	—	—	—	—	—
Fe	—	—	—	—	—	—	35.7	12.32	0.32	0.11	0.40	0.12	0.14	0.04	7.51	2.68
Cu	—	—	1.68	0.41	22.1	5.72	2.37	0.72	—	—	17.0	4.56	17.3	4.44	25.0	7.83
Zn	—	—	—	—	—	—	—	—	—	—	10.24	2.68	—	—	—	—
Sn	—	—	—	—	—	—	1.20	0.19	—	—	—	—	—	—	—	—
Pb	—	—	—	—	—	—	1.03	0.10	—	—	—	—	—	—	—	—

Table 5. The element compositions measured for the marked areas in **Figures 15, 16, 18–20.**

5.2. Results of synthesis of new particles and structures in dense deuterium using the DHPC under the action of γ-rays with 10 MeV energy

Shown in **Figure 3a** modified high-pressure chamber was used for the deuterium experiment with the initial pressure 2.2 kbar with a relative deuterium purity $\approx 10^{-6}$ (see Ref. [4]). The experiment was repeated under a pressure of 3.05 kbar with element 8 made of beryllium bronze.

The modification of the DHPC consisted in the replacement of the viton sealing with the metal-to metal one (beryllium bronze ($CuBe_2$)). This allowed to limit the chemical elements inside the high-pressure chambers filled with gases H_2, D_2 and He (HHPC, DHPC and HeHPC) only to Cu and Be. The initial pressure was constant before irradiation during 2 weeks. The atomic density $n_{D_2} \approx 2.1 \times 10^{22}$ mol.D $\cdot$ cm^{-3} and the mass density $p_{D_2} \approx 0.13$ g $\cdot$ cm^{-3} [20]. The integrated electron fluence at the target was $\approx 2.5 \times 10^{19}$ e^- during the whole experiment. After the irradiation procedure the pressure in the chamber remained constant during 1 month, so leakage of gas was not observed.

The irradiation time was $t = 1.76 \times 10^5$ s at the electron current 20–21 μA. During the irradiation at the stationary state, due to growing temperature an increase of deuterium pressure in the chamber was noted, from initial pressure $p_o \approx 2.085$ kbar at $T_0 = 293$K to $p_{stat} \approx 2.697$ kbar at $T_{stat.} = 343$ K, which means "by jump": $\Delta p \approx 612$ bar. Here, the temperature T_{stat} was measured at the surface of the pressure chamber made of beryllium bronze (**Figure 3**, position 6) inside the protecting sleeve (**Figure 3**, position 11) made from stainless steel.

Taking into consideration the known ideal gases rule [20], the calculated pressure should be: $p_{calc} = 2.441$ kbar.

Thus, the pressure change ought to be $\Delta p_{cal} = p_{calc} - p_o = 356$ bar, as opposed to direct measurement yielding the value $\Delta p_{meas} = 612$ bar. In this situation we can suppose that the temperature inside the beryllium bronze chamber was higher, attaining the value: $T_{stat,cal} = T_0\, p_{stat}/p_o \approx 379$K. That means it was 36 K higher. The interpretation of this difference can be based on the supposition that some nuclear reactions took place during our experiment, It should be noted that in the case of the experiment with hydrogen the calculated growth of pressure during the irradiation was in proper relation to the temperature measured at the surface of the pressure chamber made of beryllium bronze (**Figure 3**, position 6).

5.3. Results of measurements

An accurate SEM and XRMPA observation of the inner surface of the high-pressure chamber and inner surface of the split sleeve and closing element revealed 26 different objects which are described in a preprint published by the Joint Institute of Nuclear Research. Here we will describe only the most interesting and representative results which can be interpreted.

The object found on the inner surface of the window plug (**Figure 3**, p. 4).

Eight interesting objects were found and only two are presented in **Figure 16**. **Figure 16a** shows a curious multiple objects in the form of small pillars, which element composition is given in **Table 5**.

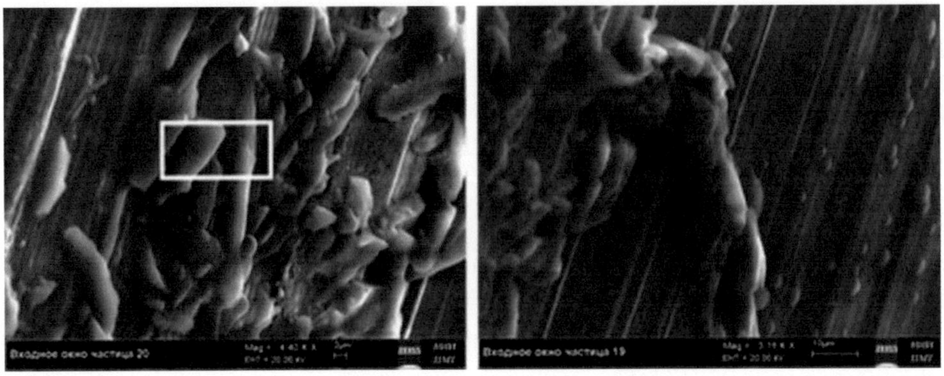

Figure 16. SEM image of the objects found on the window-plug in D_2 experiment.

5.3.1. The object found on the inside surface of the split sleeve: a container for expected products of nuclear reactions

Figure 17 gives a view of the inner surfaces of the first and second halves of the sleeve. We can see dark places on the light surfaces. The element compositions at these places are presented in **Table 6.** In the second case, we can observe a small amount of Cu, which can be interpreted as a large thickness of this object. MPRA used in our experiments collects information from the depth no greater than 4 μm.

5.3.2. The effects observed on the surface of the closing disk made of pure copper

Shown in **Figure 18a** is an image of the closing disc. One can see here similar dark stains as on the surface of the sleeve half (a). **Figure 18b** is a triangular structure on the surface of this disk with a large amount of iron (see **Table 5**).

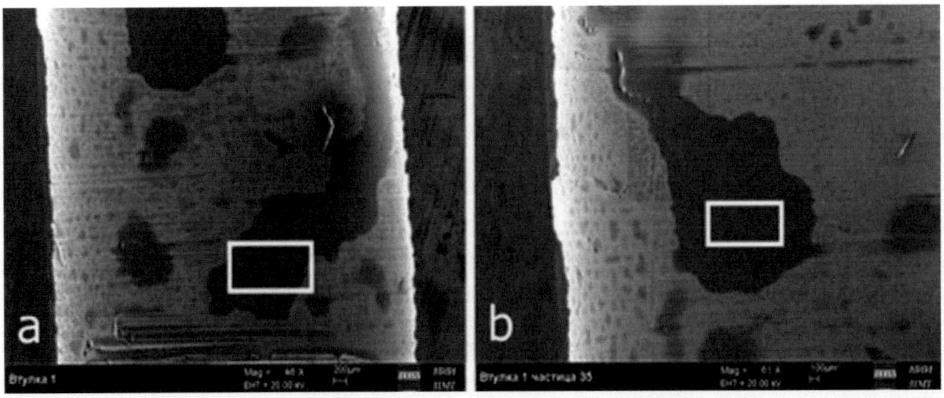

Figure 17. The observed dark stains on the inner surface of the $CuBe_2$ half sleeve, (a) and (b) in the D_2 irradiation experiment.

Element	Z	wt.%	at.%	wt.%	at.%
Area		A		B	
C	6	24.9 ± 3.9	54.92	57 ± 12	65.94
O	8	11.1 ± 1.8	18.36	39 ± 11	33.20
Si	14	0.18 ± 0.04	0.17	—	—
Cu	29	63.8 ± 1.8	26.54	3.97 ± 0.45	0.86

Table 6. The element compositions at the dark spots on the $CuBe_2$ sleeve surfaces, (a) and (b), in the marked areas from **Figure 17**.

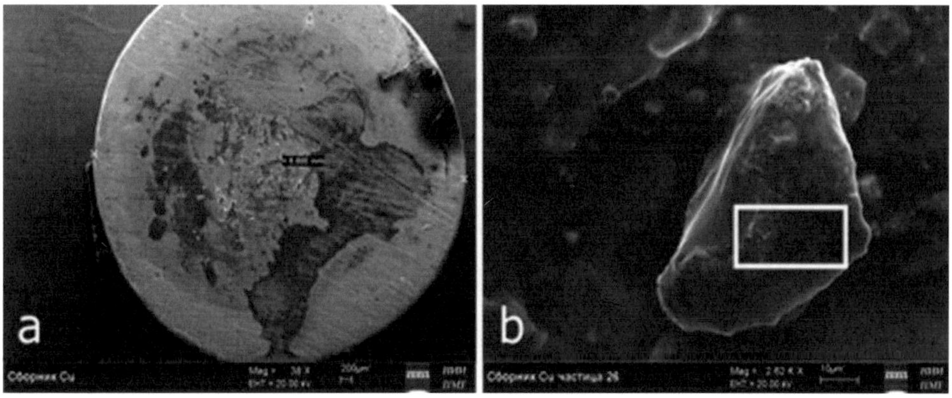

Figure 18. The view of the surface of the closing disk (a) and the found structure (b).

5.3.3. The synthesized structures found during demounting of the pressure chamber, from the left and right parts of the pressure chamber

As mentioned above, many different large particles were found after the irradiation. Here we are going to present only two typical ones. One particle, with a large amount of oxygen and of a rather regular shape, is shown in **Figure 19a**, and its element composition is given in **Table 5**. Another particle, with a large amount of carbon and of a rather non-regular shape, is shown in **Figure 19b**; and its element component is given in **Table 5**, too.

Optical Studies of the Black Oily Foils Synthesized in HeHPC under the Action of Braking γ-rays at Pressure 1.1 kbar.

Upon opening of the chamber, black foils with reinforcing needles were observed at the junction of the entrance window and reaction chamber (see **Figure 3**, pos. 5). Part of the needles protruded from the foils as lengthy white rods of uniform thickness. The foils were then placed on a clean sheet of (tracing) paper. In the place where the black foils were found, the tracing paper was soaked with exudation from the foils. Such exudation resembled synthetic oil or liquid hydrocarbons. In **Figure 20**, the photos obtained using a special device for photomicrography show an array of round black foils which acquired this shape due to the cylindrical symmetry of the reaction chamber's entrance window.

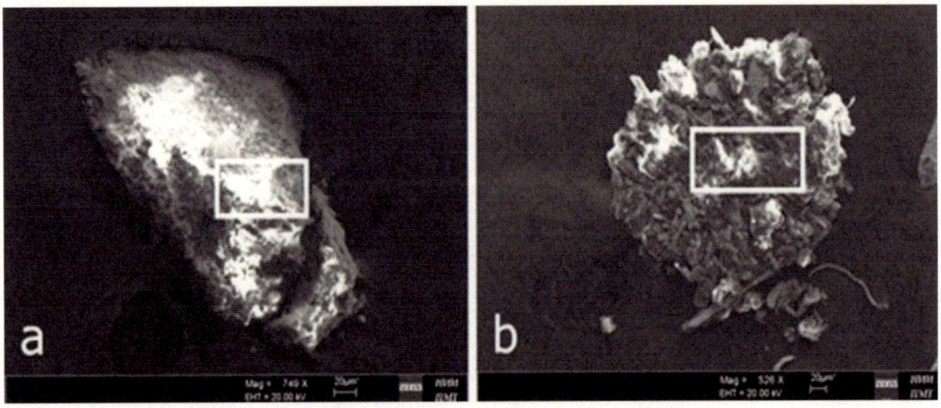

Figure 19. SEM images of the large and regular-shape particles with large amounts of oxygen (a) and with a small amount of oxygen (b) obtained in the D_2 experiment.

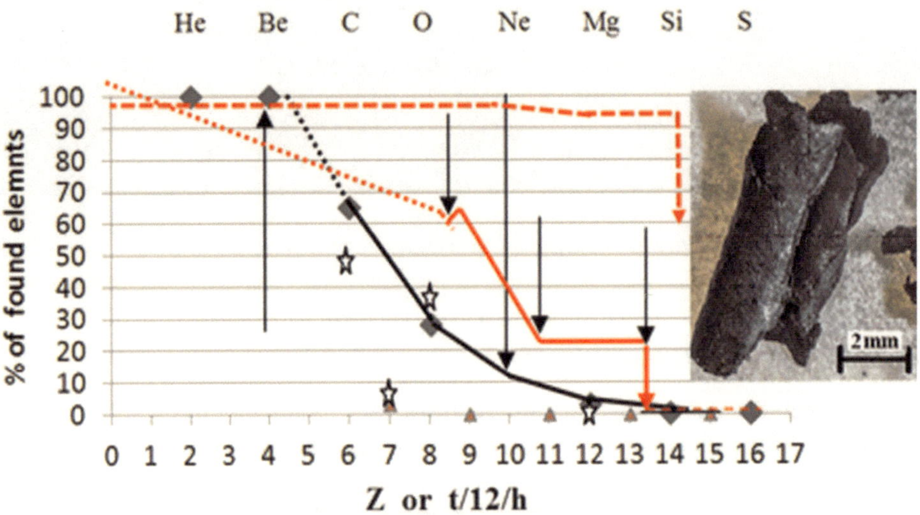

Figure 20. Illustration of the performed experiments with helium carried out using almost the same gamma irradiation procedure. Squares and the red line present first experiment data (p = 1.1 kb); stars and the dotted line correspond to the second experiment (p = 3 kb). On the right the largest object obtained in the first experiment, shown as an example.

Paper [18] (**Figures 2** and **6**) describes other black particles of complex shape obtained in the first experiment. Using the element compositions measured at some points, it was possible to determine some regularities with the help of their mean values (see **Figure 20**).

In a second helium experiment we observed another kind of inhomogeneities. Namely, the main products of synthesis of similar chemical content that covered the surface of the split sleeve surface which is presented in **Figure 21**. This experiment is described in paper [13], here only some, most interesting and representative facts are considered.

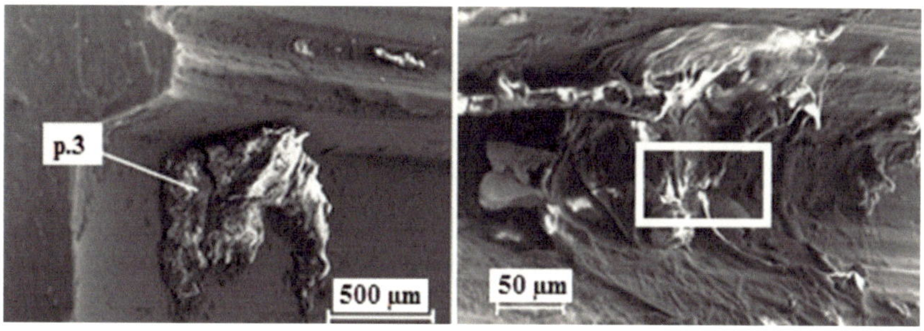

Figure 21. Large particles found near the screw surface obtained in the He experiment.

5.3.4. *The objects found on the inner surfaces of the $CuBe_2$ window-plug (**Figure 3**, position 4)*

At the surface we have observed three interesting objects, one of them being of pyramidal form. These objects are composed mainly of carbon (26.5 at.%), oxygen (62.31 at.%) and potassium (8.33 at.%).

5.3.5. *The objects observed on the $Cu_{0.98}Be_{0.02}$ cylinder*

On the surface of the cylinder (divided into two equal parts) we could observe 27 different objects. Three of them, of large dimensions and complicated form, located on the wall, are presented in **Figure 21** and their element compositions, in p. 3 and rectangular, are given in **Table 5**.

An example of deposition on a wall and an object of complicated form are shown in **Figure 22**. In the middle, there is a photo of the respective surface before irradiation.

Figure 20 summarizes the element compositions measured for different objects. The curves are given for two experiments with account of the mean values for the element compositions obtained using the MPRA method. Presented curves are different but have a similar character, i.e., a regular decreasing tendency of atomic concentration with a rising atomic number. In

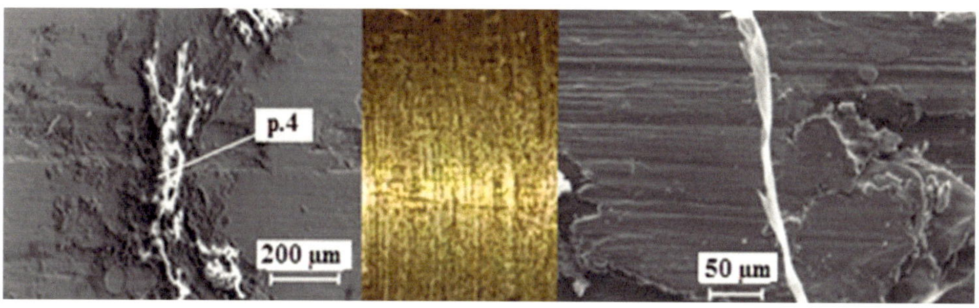

Figure 22. SEM images of the inside surface of the $CuBe_2$ sleeve. Here, we can see specific surface structures and a particle of complex shape. The chemical composition of the particle at p. 4 (in at.%) is as follows: C—39.58, N—9.99, O—39.35, Na—0.69, Cl—1.49, Ca—0.32 and Cu—6.69. In the middle, there is a photo of the respective inside surface of the sleeve prior to irradiation in the second He experiment.

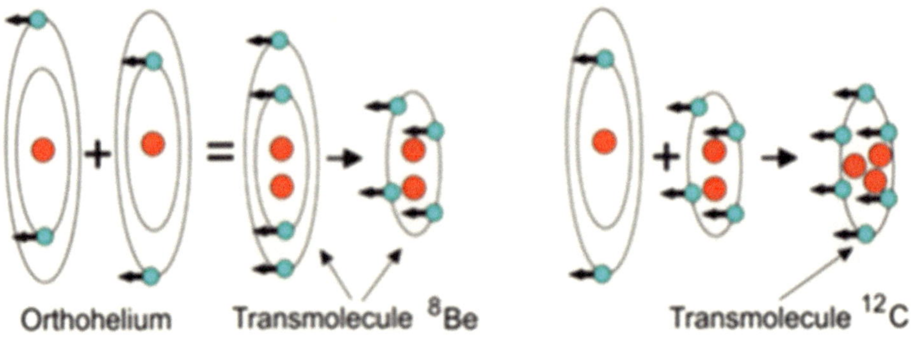

Figure 23. Creation of the stabilized transmolecule of the Carbon $^{12}C_{tr}$. Red point means the α particle [41, 42].

those experiments, inside the HeHPC there was either spectrally pure helium under different pressures or a palladium rod in such helium atmosphere.

All of the three synthesized particles here are the largest objects shown in **Figures 21** and **22** (see also Refs. [12, 13]), and have dielectric properties. According to the X-ray structural analysis, these articles have amorphous structures with a complex hydrocarbon composition that is absent in the reference library of the X-ray analysis. Due to its importance and for the purpose of greater reliability of the data, the structure of synthesized particles was analyzed using microscopes and X-ray micro-analyzers at two independent analytical centers. The mechanical properties of the large particles were not investigated. But there is a most intriguing problem with the electrical properties of the carbon-based structures which we obtained during all the helium experiments. The latest measurements of object shown in **Figure 22** (right upper corner) [30] have shown a very large electrical resistivity $\rho = 10^5 \mu\Omega m$, relative dielectric constant ε = 3–4, low density d = (1.2 $\pm$ 0.2)g/cm^3, appeared large paramagnetic properties, having main chemical content C_{62}, O_{31}, Mn_3 wt.%, structure-open problem. In the compression probe we have noted its micro elastic-brittle state with strength 10 MPa (approximate value) and temperature stability bellow 400°C. To interpret these results, phenomenological model of deuteron photo-disintegration with Oppenheimer's reaction, atomic processes of γ-ray scattering on atoms, formation of a di-nuclear system, quasi-classical approach with a two-hump ion potential, internal conversion in solids using a third charged particle, stabilization of nuclei in respect to the β-decay with occupied electronic shells, interpretation of cluster radioactivity, synthesis of elements in astrophysics and element abundance in the Universe were tested in other publications [21–29]. The induced effect in case of helium, referred to multi nuclear reactions in condensed (under high pressure) helium. This process could be considered as cold transmutation of nuclei of chemical elements phenomena, where the unusual detection of transmolecule 12C could be explained as seen in **Figure 23** [30].

6. Conclusions

Induced effects of gamma irradiation at different energies (10 and 23 MeV) were presented in different pure elements and in more complicated systems in details, the results indicated "upstream" nuclear reactions with synthesis of heavier elements and "downstream" reactions with asymmetric fission, with the formation of both light elements from carbon to iron and heavier elements (Ag, Zr, Ba, Ta, W, Pb, Pt, Au) were formed. To interpret these results several models were used. From the authors' point of view, to achieve an accurate description of these effects it is important to elaborate on new approaches to fission and fusion nuclear reactions. Because of volume limit of the chapter, problems including how to explain "protuberation," macro mass changes, so called "Dubna Oscillation," [31] how to realize third step of Tsar bomb (using it for public welfare) author decided to publish it later.

Acknowledgements

Many thanks to Teresa Wilczyńska-Kitowska for her close, long time and fruitful cooperation.

All experiments were carried out according to International (PL-RF) Scientific Cooperation in very efficient, friendly atmosphere, during many years of close cooperation between Polish and Russian teams (presented by Author and Alexander Yu. Didyk) and were financed by both sides.

Author details

Roland Wiśniewski[1,2,3]*

*Address all correspondence to: roland.wisniewski@gmail.com

1 Joint Institute for Nuclear Research, Dubna, Russia

2 National Centre of Nuclear Research, Otwock-Świerk, Poland

3 Physics Faculty, Warsaw University of Technology, Warsaw, Poland

References

[1] Yu DA, Wiśniewski R, Wilczyńska T. Joint Protocol JINR-IAE 2009-2011 and JINR-NCNR 2012-2014. Poland: NCNR; Archives. 2013

[2] Baranowski B, Wiśniewski R. Synteza wodorku niklu z metalicznego niklu i gazowego wodoru. Bulletin de l'Academie Polonaise des Sciences, Serie des Sciences Chimiques. 1966;**14**(4):1273-1275

[3] Wiśniewski R. High pressure apparatus for Gaseous Hydrogen (Deuterium) up to 25 kilobar and temperature range −50°C – +100°C. Review of Scientific Instruments. 1970;**41**(3):455-464

[4] Didyk AY, Wiśniewski R. Nuclear reaction, induced by γ quanta, in palladium saturated with deuterium saturated surrounded by dense deuterium gas. Europhysics Letters. 2012; **99**:22001-22006

[5] Didyk AY, Wiśniewski R. Nuclear Reactions in deuterium-saturated palladium under irradiation by 10MeV γ quanta in dense molecular deuterium at 1.2kbar pressure. Europhysics Letters. 2013;**103**:42002-42008

[6] Didyk AY, Wiśniewski R. Synthesis of micro particles in molecular hydrogen at 1 kbar pressure in nuclear reactions induced by braking γ rays of 10MeV threshold energy. The Chemical Composition and Structures at the Inner Surfaces of the Pressure Chamber Components. Preprint P15-2014-2, Dubna: JINR, RF, 2014. pp 41

[7] Didyk AY, Wiśniewski R. Nuclear reactions in palladium saturated with deuterium and rhenium in dense deuterium gas under irradiation by γ quanta of continuous spectrum with boundary 23MeV energy. Preprint R15-2012-63, Dubna: JINR, 2012. pp. 21

[8] Didyk AY, Wiśniewski R. Results of radiation of aluminum and homogeneous alloy YMn2 by 23MeV energy γ quanta in the atmosphere of molecular deuterium at 2kbar pressure. Preprint P15-2013-41, Dubna, JINR, RF, 2013. pp 18

[9] Didyk AY, Wiśniewski R. Changes observed in the surfaces, bulk properties and chemical composition of vanadium and stainless steel specimens irradiated in dense gaseous deuterium by γ quanta of threshold energy 23 MeV. Preprint P15-2012-75, Dubna: JINR, RF, 2012. pp 14

[10] Didyk AY, Wiśniewski R. Results of radiation of complex system –Al-Y-Mn2- Al- Y-Mn2-Cu- SS by 23MeV energy γ quanta in the atmosphere of molecular deuterium at 2kbar pressure. Preprint P15-2013-41, Dubna: JINR, RF, 2013. pp 19

[11] Didyk AY, Wiśniewski R. Results of radiation of complex system Sn- Mo- Fe-Ni-Bi-Ta-Cu by 23MeV energy γ quanta in the atmosphere of molecular deuterium at 2kbar pressure. To be published

[12] Didyk AY, Wiśniewski R, Wilczyńska-Kitowska T. The carbon-based structures synthetized through nuclear reaction in helium at 1.1kbar pressure under irradiation with braking γ rays of 10mev threshold energy. Europhysics Letters. 2015;**109**:22001-22006

[13] Didyk AY, Wiśniewski R. Synthesis of new structures and substances in dense gases H2, D2, and He under irradiation by braking 10MeV γ rays in $Cu_{0.98}$ $Be_{0.02}$ pressure chamber. JPSA. 2016;**6**(4):13-21

[14] Didyk AY, Wiśniewski R, Wilczyńska-Kitowska, T. Some physical properties of graphite-like object obtained by γ quanta irradiation with threshold energy of 10MeV of pure gaseous he under pressure, in CuBe2 apparatus. JPSA. 2016;**6**(6):1-11

[15] Didyk AY, Wiśniewski R. Synthesis of micro particles in molecular hydrogen at 1kbar pressure in nuclear reactions induced by braking γ rays of 10MeV threshold energy. The chemical composition and structures at the inner surfaces of the pressure chamber components. P15-2014 – 2, pp. 40. Particles and Nuclei, Letters. 2015;**12**(7):1298-1317

[16] Didyk AY, Wiśniewski R, Myshinsky GW, Semin WA, Wilczyńska-Kitowska T. Synthesis of solid-state structures and chemical elements under irradiation by bremsstrahlung γ

rays with maximum energy of 10MeV in condensed deuterium at a pressure of 2.2 kbar. Preprint P15-2018- 3. Dubna: JINR, RF, Particles and Nuclei, Letters 2018, in press. 2018. pp 20

[17] Ishkhanov BA, Kapitonov IM. Interaction of electromagnetic interaction with atomic nuclei. MSU. 1979:215

[18] Bethe H, Morisson P. Elementary Nuclear Theory. New York: John Willey & Sons, Inc; 1956. p. 356

[19] Mukhin KN. Experimental nuclear physics. Physics of Atomic Nucleus. Moscow: Energoatomiz-dat. 1983;**1**

[20] Didyk AY, Wiśniewski R. Properties of hydrogen and its isotopes under high pressure and technological applications. Dubna: JINR. 2013. pp. 320. ISBN: 978-5-9530-0358-2

[21] Sierk AJ. Mass-assimetric fission of light nuclei 1985. Macroscopic Model of Rotating Nuclei. Physics Review. 1986:**55**(6):582-583

[22] Oppenheimer JR, Fillips M. Note for the transmission functions for deuteron. Physics Review. 1935;**48**(15):500-502

[23] Tarakanov AV, Shilov VM, Schmidt R. Nuclear Physics. 1991;**53**(5):1285-1291

[24] Shilov VM. Sub-barrier fusion of spherical nuclei of medium atomic number. Nuclear Physics. 2012;**75**(4):485-490

[25] Kálmán P, Keszthelyi T. Lattice effects in solid state internal conversion. Nuclear Physics. C79. 2009;**031602**(R):P.031602-1-031606-4

[26] Bosch F, Faestermann T, Friese J, et al. Observation of bound-state β- decay of fully ionized ^{187}Re: ^{187}Re - ^{187}Os Cosmochronometry Decay. Physical Review Letters. 1996;**77**(26):5190-5193

[27] Volkov V V, Cherepanov EA. Formation of nuclear molecules in cluster radioactivity. On interpretation of the cluster radioactivity mechanism. Physics of Particles and Nuclei Letters. 2013;**3**(180):347-353

[28] Ishkhanov BS, Kapitonov IM, Tutyn IAM. LIBROKOM. 2009. http://nuclphys.sinp.msu.ru/nuclsynt.html

[29] Bethe HA. Energy production in stars. Physics Review. 1939;**55**:434-456

[30] Mishinsky GV. Multinuclear reaction in condensed helium. Nuclear Physics. 2017;**9**(1):94-105. RENSIT and Nuclear Physics, RENSIT. Atom in a strong magnetic field. Transformation of atoms into transatoms. 2017;**9**(2):147-160

[31] Wiśniewski R, Wilczyńska-Kitowska T, Didyk AY. Dubna oscillations phenomenon. In preparation for publication

Section 3

Nuclear Thermal Hydraulics

Chapter 4

Dense Granular Flow as Heat Transfer Media: A New Type of High Power Target Design

Xuezhi Zhang and Lei Yang

Additional information is available at the end of the chapter

http://dx.doi.org/10.5772/intechopen.77276

Abstract

High power target systems require reliable high-temperature heat transfer media. Dense granular flow materials have potential to work as heat transfer media, especially at high temperature. In this chapter, a brief review of dense granular and their heat transfer properties is introduced, including basic concepts of heat transfer, thermal behavior of these materials, and factors that affect this behavior. The implementation of these materials as targets is addressed, where two targets were designed, constructed, and tested based on the concept of dense granular flow. The results of the application of a hopper flow-type target in 2.5 MW accelerator-driven system (ADS) and a chute flow-type target in the material irradiation facility will be presented.

Keywords: dense granular flow, high power target, prototype facility

1. Introduction

In nuclear area, an accelerator is always helpful to generate particles uncommon or indiscrete in the natural world. Target materials are exposed to certain projectiles; as a result, secondary particles in the form of neutrons, photons, neutrino, and so on are generated.

For most cases, the higher particle flux outcome is more helpful. To achieve this, the target systems are designed to stand high beam intensity. Most of beam power deposits in the target material. The kinetic energy of the beam converts into internal energy and increase the temperature of target. This process also leads to radiation damage of target material and even target failure.

Lots of effort has been made to achieve high power target. But not until the 1990s, a specific megawatt target never came up. As examples, Weapons Neutron Research project in Los Alamos Neutron Science Center achieved 70 kW target with tungsten material, while Rutherford Appleton Laboratory reached 160 kW using 800 MeV protons and tungsten with tantalum cladding [1].

As mentioned above, cooling is essential to the high power target. Conventional cooling channels take too much space inside the target body and gradually get close to the limit. The MarkIII target of Swiss Spallation Neutron Source project in Paul Scherrer Institute (PSI) reaches 1.2 MW beam power based on such concept. Some projects adopt rotating target to even and lower the power density within the target body, such as European Spallation Source and China Spallation Neutron Source [1–4].

For solid target, radiation material damage also causes swelling and creep, which might damage the cooling channel and cause cooling deterioration, as the displacement per atom (DPA) increases. One way to bypass the damage is using liquid metal as target material. Meanwhile, the flow of liquid metal removes the power deposit and dumps the heat in the heat exchanger. These properties make liquid metal target a promising solution of high power target. In PSI, MEGAwatt PIlot Experiment (MEGAPIE) achieved 0.8 MW with lead bismuth eutectic (LBE). Spallation Neutron Source in US and Japan Spallation Neutron Source also achieved 1.4 and 1 MW beam power with mercury [1].

Here we are going to present a third option of target cooling, which is based on the dense granular flow [5]. Granular system makes solid materials flow like fluid and perform heat removing. Meanwhile, the candidate target materials are more various, which give better feasibility depending on the purpose of the facility.

2. Heat transfer media and heat transfer in granular material

2.1. Conventional heat transfer media

The continuous research of the thermal properties of materials is explained both by practical needs and fundamental science. Heat removal is always a crucial issue for energy conversion in the energy industry owing to increased levels of power requirement and energy efficiency. The search for materials that transfer heat well under high temperature and energy flux has covered all the industry area and never ends, and it is especially essential for design of the next generation of reactors and neutron facilities.

A medium's transfer heat ability follows three basic heat transfer ways, conduction, radiation, and convection. Conduction dominates solids' heat transfer behavior, while convection makes fluids absorb heat from a source and dump heat to a sink very efficiently. In field of nuclear energy, water is still the most important heat transfer material for cost and reliability reasons, while liquid metal such as sodium and gases such as helium take the rest portion. These fluid heat transfer media are capable of removing 10^2–10^5W/m^{2}K from surface of power component and reaching up to the order of ~10^6W/m^{2}K while boiling [6]. At high temperature, some particular situations differ because of radiation. Radiation transfers heat from one surface to another through void following the Stephan-Boltzmann law, which indicates the radiation heat transfer rate is proportional to the cubic of temperature. Hence, if radiation happens effectively in the media, heat transfer can be more efficient at high temperature.

At high temperature, thermodynamics principles allow fluid's thermal energy transfers more efficiently. But most practical heat transfer media failed under such temperature: Liquid water transforms to supercritical state over 385°C [7, 8]; steam reacts with common structural materials over 900°C [9], and about one-third of the zircaloy in the core create a sizable hydrogen bubble from the Zr-steam interactions [10]; liquid sodium or lithium, Na-K alloy, and mercury are all vaporized under 900°C; lead-bismuth eutectic, liquid lead's corrosion-erosion effect grows significant at 480–550°C [11, 12]. What's more, for nuclear usage, liquid metal loop needs multiple conditioning systems, which makes the system complicated and expensive, such as International Fusion Materials Irradiation Facility (IFMIF) [13] and eXperimental Accelerator-Driven System (XT-ADS) device [14]. Gases seem to be another good choice, which has been testified in high-temperature gas-cooled reactors. But they have to be pressurized due to their low volume heat capacity [15].

2.2. Characteristic of granular flow

Granular material is a promising solution. Granular material is composed of discrete solid particles and gaseous environment where the particles immerse. Such material has been used for heating since ancient Rome by Lucretius [16]. Modern research can be trace back to Maxwell's era and got broad attentions in the twentieth century due to popularization of bed reactors [17]. Nowadays, granular heat exchanger has been widely used in many industrial areas. They perform effective heating or cooling in so many industrial processes that indicate granular materials are realistic to be a promising heat transfer media. Packed bed and powder jet solution had been came up by research groups to solve the high power density removal issues of target [18, 19]. Pebble bed fusion reactors and fission first wall breeding layers also adopt granular material using the unique advantages in the high temperature heat transfer [20–23]. Flow of granular state enhances heat transport process, leading to better heat transfer performance and more realistic feasibility, especially for extreme internal heat source such as spallation reaction, compared to present solutions.

Its potential of heat transfer usage stems from flowability and thermal properties, which has been widely studied and practiced. The flowability allows granular materials to convey long distance by multiple methods. But granular materials' flowability is quite different from fluids'. Although granular dynamics uses many hydrodynamics concepts, granular dynamics behaves quite different from fluid. Some of granular materials' unique properties are critically helpful to be heat transfer media.

When granular materials flow out from a vessel under the action of gravity, the mass flow rate M is approximately independent of the head of material head H, just as an hourglass does. The most important research of this phenomenon was successively recorded by Beverloo et al., Aldin and Gunn, and Nedderman [24–28]. This phenomenon requires that the granular material must be not too smooth and small, while the bulk has a height not less than twice orifice diameter. A quantitative result of this flow gives a relationship of [28].

$$\dot{m} = 0.58\,\rho_s \sqrt{g}\,(D - 1.4\,d_p)^{2.5}\,F(\theta_\omega, \beta) \quad (1)$$

where m, ϱ_s, D, d_p, θ, and β represent mass flow rate, particle material density, orifice diameter, particle diameter, wall angle, and an angle related to friction of the material [29]. This result implies the flow rate could be controlled by local valve.

As the immerging depth increase, the normal pressure of most fluids increases linearly and is proportional to the depth. But the normal pressure of granular system gradually reaches a constant. Different from fluids, inside the granular system and between the granules and the container walls, friction exists. The upward friction force exerted by the wall neutralizes the weight of grains. An important quantitative explanation given by Janssen is in accord with the experimental stress result [30].

Other granular properties also have special effects, which stem from the friction, dissipation, and size effect of granules. These effects lead to phenomenons in granular flow, such as the jamming and dilatancy properties prevent flow channel collapse [31]; while dissipation property prevents long-range disturbance propagation [32].

2.3. Thermal properties of granular flow

Granular materials' thermal properties are decided by the components which make the effective properties vary largely. As high-temperature heat transfer media, high-melting-point metal alloy, ceramics, and other materials with good heat conductivity can be the solid particle candidate. There are relatively few gases performing heat transfer effectively, among which helium is representative. The large number of material combination makes granular media satisfy varieties of actual needs. For instance, the heat transfer media in traditional nuclear reactor need to minimize the neutron absorption and maximize the neutron moderation, while the heat transfer media had better to be the spallation material in a spallation target which needs high spallation neutron production.

Some effective thermophysical properties such as critical temperature, density, and specific heat can be estimated using simple calculations based on the properties of the initial components. But quantities such as heat transfer coefficient and effective heat conductivity are far more complicated than those. The following subsections will review critical thermal properties to get a better view of granular heat transfer media's potential.

The mainstream viewpoint of heat transfer between granular material and a panel has two stages: heat transfers from the wall surface to the first granular layer next to the wall, and heat diffuses from the first layer to the inside of granular material [33–36]. Sullivan and Sabersky [36] adopted concepts of fluid heat transfer such as Peclet and Nusselt number dealing with this process. An analytical result of temperature distribution can be derived from principles of heat transfer. This work was further extended by Spelt et al. [36, 37].

2.3.1. Mechanisms

For both two stages, whatever discrete model or continuous model is applied, one should realize that the structure of granular material leads to more heat transfer mechanisms.

As Yagi and Kunii reviewed, heat transfer in granular media happens in all three ways and their coupled effects [38]. Most concerned mechanisms of heat transfer in granular media are listed in **Table 1**.

The first stage was observed by several early researchers, depicted as a sharp temperature drop at the wall [36, 39, 40]. In this stage, the maximum heat transfer is at the order of ~10^3W/m^2K for granules' diameter at the order of millimeter in air. Such order is determined by different researchers and methods [39, 41]. When the pressure drops to 0.1 Pa, the wall heat transfer coefficient decreases to the order of ~10 W/m^2 K. Most significant heat transfer coefficient drop happens when pressure changes from ~1 kPa to 1 Pa, leading to the change of one and a half order.

In vacuum, the solid dominates the heat transfer rate, proved by the coefficient of bronze and glass, which are 30 and 5 W/m^2 K, respectively. This result agrees well with the contact heat transfer analysis, and we will discuss this effect in the effective thermal conductivity of granular material in later section. In contrast, the gas phase decides the heat transfer in atmosphere, reaching the order of ~10^3W/m^2K [35, 41]. It is believed that the porosity near the wall reaches almost 1 [42–45], which means only little solid contact at the wall leads to a minor heat transfer mechanism only significant in vacuum.

Although most researches did not evaluate the specific limit of heat transfer rate, they did observe effect of various factors. Most of them have relationship with the second stage.

	Phase	Region	Mechanism
1	Solid	In single particle	Conduction
2	Solid-Solid	Contact	Conduction
3	Solid-Solid	Surface to surface	Radiation
4	Void-Void		Radiation
5	Solid-air-Solid	Near the contact	Conduction
6	Solid-air-Solid		Convection
7	Air-Air	Long range mixing	Convection
8	Air	In the void	Conduction

Table 1. Mechanism of heat transfer in granular media.

In this stage the heat transport can be considered as heat diffusion in a continuous media with effective properties of granular material. The complexity of heat transfer mechanism makes multiple parameters dependence.

2.3.2. *Effect of material constituent*

Granular material component combination is the most principle factor of heat transfer properties. In the past century, a number of various granules' heat transfer properties have been studied. A collection of experimental data is shown in **Table 2**. It shows gas phase plays a decisive role in granular materials' heat transfer property, while the solid particle geometry plays a role as important as its inherent properties. Large surface-volume ratio makes the temperature of fluid and solid phase nearly uniform [46, 47].

In energy area, most heat transfer is performed under pressure. Pressurization enhanced heat transfer in two aspects: increase boiling temperature and volume heat capacity. High-melting-point solid granules need none of these improvements and hence may significantly reduce engineering difficulty, risk and cost. The pressure concept in granular system also has two aspects: the gas pressure and particle compress pressure.

Early stage of ETC of granular material focuses on experimental determination. Back to 1926, Aberdeen and Laby studied effective thermal conductivity (ETC) of monox powder when immersed in air, in carbon dioxide, and in hydrogen at various pressures has been determined. They found a linear relation between the conductivity of the powder and the logarithm of the pressure of the gas in which it is immersed. The result fits in

$$k = \frac{k_f}{2}\log_{10}\frac{p}{n} \qquad (2)$$

where k is effective heat transfer coefficient and k_f is fluid heat transfer coefficient and n is a constant for each specific gas [48].

2.3.3. *Effect of temperature*

Microscope heat transfer mechanism decides most condensed matters' thermal conductivity deteriorates as temperature increases while gases are contrary. In a granular material, solid phase has more significant heat conductivity than gas phase. On the other hand, the packed solid particle structure is point contact for most cases; hence, the gas voids play an important role in the effective heat conduct which increases as temperature rises. Radiation also significantly improves effective heat conductivity of granular material under high temperature. Solid surfaces and voids allow radiation to perform efficient heat transfer, while such mechanism does not exist in bulk solid or pure gas.

Yagi and Kunii [38] determined ETC for several materials, in the temperature range of 100–800°C. The result shows an accelerated increase of ETC as the temperature increases. Most materials' ETC doubled or tripled for the test range. They proposed a model considering

Solid	Diam./mm	Gas	Pressure/Pa	Temp./C	Effective Heat Transfer Coefficient/ W/m²K
Glass	4	Air	101k	60	0.28
Al	3.18	Air	101k	300	0.11
Iron	11	Air	101k	200	0.46
Iron	11	Air	101k	500	0.69
Iron	11	Air	101k	700	0.92
Iron	11	Air	101k	800	1.08
Glass	2.6	Air	101k	65	0.17
Steel	4.8	Air	101k	65	0.29
SiC	0.55	H_2	150		0.48
SiC	0.55	H_2	101k		1.06
SiC	0.55	CO_2	170		0.11
SiC	0.55	CO_2	101k		1.06
SiC	0.55	Air	150		0.13
SiC	0.55	Air	101k		0.23
Glass	0.32	He	76k		0.36
Glass	0.32	He	130		0.15
Glass	0.32	Air	101k		0.16
Glass	0.32	Air	430		0.12

Table 2. Some experimental result of effective heat transfer coefficient (ETC) [39, 49, 50].

the effect of radiation [38]. As Botterill published in 1989, the radiation is proportional to T^3, σ, and the character length represented by D_p; hence a radiation term of ETC falls in the form of

$$k_r = 4\sigma\chi D_p T^3 \tag{3}$$

where σ is Stefan-Boltzmann coefficient and χ is coefficient depending on the specific model. A comparative research of various radiation models and experimental results was also given by Botterill [49].

In a word, unlike traditional heat transfer material, granular material could perform better under high temperature; as previous study of gas cooling reactor, heat conductivity of graphite pebble bed exceeds pure graphite for temperature higher than 1400°C [2]. Further experimental results were reported in [38, 49, 50].

2.3.4. *Effect of gas motion*

As is well known, forced convection enhanced heat transfer properties. Packed bed is motionless granules with gas flows through. Researches of these systems explain the motion effect of gas phase, which also implies the same effect in granular flow.

Bunnell et al. studied a wall-heated packed bed with different superficial gas velocity, measured the inner temperature distribution, and calculated ETC through these results. They found effective thermal conductivity dependence on mass velocity of the gas, and the gas phase and solid phase temperatures are nearly uniform [51].

Solid particle motion enhanced cold and hot particles mixing, hence increasing the heat transfer rate. A lot of research about packed bed also showed that only fluid flow rate increase also leads to enhancement of ETC [46, 50, 52–55].

2.3.5. *Models of ETC*

To get a quick evaluation of granular materials, researchers had come up with various models, from simple to complex. Three most important models are as follows:

Maxwell (1881):

$$k = k_g\left[\frac{2\varepsilon k_g + (3 - 2\varepsilon) k_s}{(3 - \varepsilon) k_g + \varepsilon k_s}\right] \tag{4}$$

This model is probably the earliest method for granular ETC evaluation. It represents a series model that considers the system as a parallel-serial thermal resistance system [53] which got its popularity in the early age.

Yagi and Kunii [38]:

$$k = k_g\left(\frac{\beta(1 - \varepsilon)}{\gamma\left(\frac{k_g}{k_s}\right) + \frac{1}{\frac{1}{\varphi} + (D_p h_{rs}/k_g)}} + \varepsilon\beta D_p h_{rv}/k_g\right) \tag{5}$$

in which $\beta = l_p/D_p$, $\gamma = l_s/D_p$, $\varphi = l_v/D_p$. h_{rs}, h_{rv}, l_p, l_s, and l_v represent radiation heat transfer coefficient of solid to solid, void to void, effective length of centers, thermal conduction, and fluid film at the contact point. They considered effect of fluid flow using Ranz fluid mixing equation and shows the particle size effect. The result fits temperature effect well [38].

Zehner and Schlunder's model:

$$k = k_f(1 - 1\sqrt{1-\varepsilon})\left(1 + \frac{k_r}{k_f}\right) + k_f\sqrt{1-\varepsilon}\left\{\frac{2}{1 - \frac{k_f}{k_s}B}\left[\frac{\left(1 - \frac{k_f}{k_s}\right)B}{\left(1 - \frac{k_f}{k_s}B\right)^2} \cdot \ln\frac{k_s}{k_f B} - \frac{B+1}{2} - \frac{B-1}{1 - \frac{k_f}{k_s}B}\right] + \frac{1}{\frac{k_f}{k_r} + \frac{k_f}{k_s}}\right\} \quad (6)$$

in which B is a deformation factor. This model represents a significant improvement of contact term which applies for various material properties.

2.3.6. Total heat transfer coefficient

Schlunder gave explanation of the combination of wall heat transfer and ETC as a serial of heat resistance as follows [35]:

$$\frac{1}{\alpha} = \frac{1}{\alpha_w} + \frac{1}{\alpha_{ETC}} \quad (7)$$

Because the heat diffusive process inside the granules varies as time changes, total heat transfer coefficient also depends on time. Most measured total heat transfer coefficient is time averaged value:

$$\alpha = \frac{1}{t}\int_0^t a_t\, d\,t' \quad (8)$$

In three situations, (1) t goes to 0, (2) bed heat capacity is infinity, and (3) the bed is under perfect mixing, α_{ETC} reaches a maximum and hence derives the maximum heat transfer coefficient, i.e., the wall heat transfer coefficient. A typical time dependence of α_{ETC} based on Fourier theory is given by

$$\alpha_{ETC} = \frac{2\sqrt{(\rho c k)_{bed}}}{\sqrt{\pi t}} \quad (9)$$

The total heat transfer depends on time, hence given by.

$$\alpha = \alpha_w \frac{1}{1 + \frac{\sqrt{\pi}}{2}\sqrt{\tau}}, \quad \text{where} \quad \tau = \frac{\alpha_w^2}{(\rho c k)_{bed}} t \quad (10)$$

Through the experiments of various components' combination, Ernst showed α decreases as time increases, which is less significant for lower pressure and larger granules [39]. Such time-dependent phenomenon has been observed by other researches [36, 56]. The total heat transfer coefficient fell in the range of 50–200 W/m²·K which is also a typical range of most facility situation [57].

3. Implementation of granular flow target

With the concept of dense granular flow cooling, the target can be designed according to different demands. Two different types of granular flow target have been developed for China Initiative Accelerator-Driven System (CIADS) project and China material irradiation facility project. For both designs helium was selected as the cover gas, concerning its high thermal conductivity and safety. In principle, pressure was set lower than 1 atm to prevent leakage and help target-beam line coupling. For each case, gas pressure was variable parameter. The difference of the two targets includes structure, target material, beam energy, spectrum, and so on.

By the end of 2017, two prototype facilities were constructed and tested with electrical heating, up to about 300°C. The beam coupling experiments are planned to be performed later, depending on the progress of radiation shielding hall.

3.1. ADS target

For CIADS, the target needs to prove high neutron yield inside a subcritical reactor core. To replenish the neutrons in a subcritical power reactor, the target-beam power is very high. The primary design chose proton beam of 500 MeV @ 5 mA and beam power of 2.5 MW.

Meanwhile, the space for the target installation is very limited to lower the neutron leakage and increase the effective neutron multiplication factor (k_{eff}) of the reactor. Usually the installation space is in vacancy of several removed fuel components. In CIADS, the diameter of the target tube needs to be under 30 cm. To fit this space, a thin and long vertical tube was applied as target tube. The dense granular flow went downward within. The beam line installed coaxially in the target tube sent protons deep into the target granules and reacted at the same level of the core center. **Figure 1** shows a schematic view of CIADS, which adopts a loop target section.

Tungsten was selected as the target material. Minor iron was added to optimize the mechanical properties. The alloy was made into 1 mm sphere. Tungsten generates abundant neutrons in spallation reaction which is widely adopted in neutron resources. With the basic idea of dense granular flow, the tungsten also plays a role as coolant. Considering the beam heat load and material's temperature endurance, the circulation mass flow is required about 200 kg/s. Compared to the fluidized dilute granular flow developed in the Rutherford Appleton Laboratory, we believed that the dense flow is more stable and lowers the requirement of gas system, for this specific case.

To sustain the operation, the system also included conveying, cooling, conditioning, monitoring, cleaning, and other affiliate systems. To lift the spheres back into the target, a chain conveyor is employed. The spheres inside flow vertically under gravity. When the spheres

http://dx.doi.org/10.5772/intechopen.77276

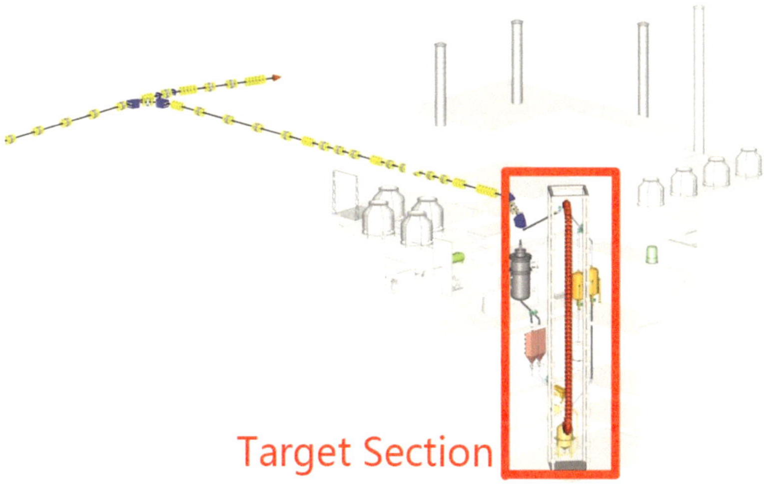

Figure 1. Schematic diagram of CIADS.

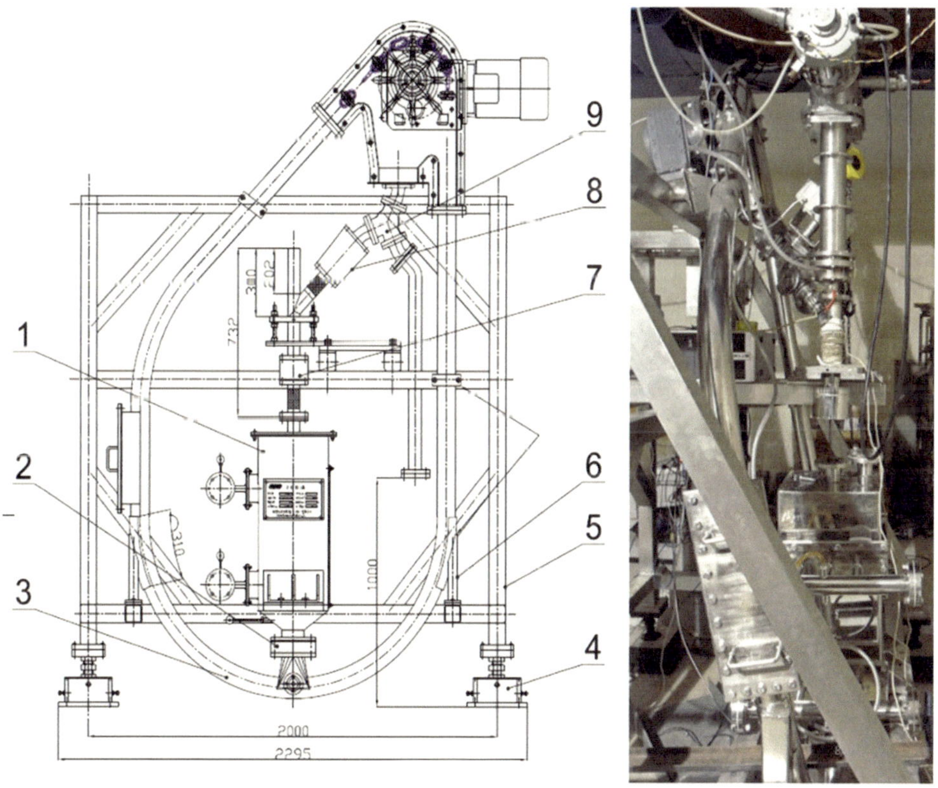

1. Heat exchanger 2. Main valve 3.Tube chain conveyer 4. Pedestal
5/6. Frame 7.Beam coupling section 8. Flowmeter 9.Switch

Figure 2. Electron beam coupling test facility.

flow to the bottom of the loop, the elevator lifts them back to the top, and the spheres reenter into the target. The elevator actively supplies the energy to sustain the circulation; hence the reliability and feasibility should be carefully considered. Other confines came from the ADS application, such as airtight and compact. Due to the grain size and density, pneumatic system does not fit the system well. Both conventional mechanical conveyer and a new developed magnetic conveyer are chosen as potential candidate. The heat of spheres will be removed in a panel heat exchanger. Through the steel shell of the heat transfer panel, the heat is transferred to the counter water flow. The width of parallel flow channel formed by the panels is large enough to avoid arching and clogging. **Figure 2** shows an electron beam coupling test facility constructed in 2015. Without conditioning and some other minor systems, the facility is simplified to test the circulation and heat loading effect. As shown in **Figure 3**, the coupling section is about 1:5 to the full scale.

The full-scale testing was setup later in 2017, as shown in **Figure 4**. The model facilities were put into practice as a preliminary design verification by Institute of Modern Physics, Lanzhou. Other than the components of previous system, sieving and cover gas systems were installed. Conveying system was updated to a magnet conveyer, and heat exchange test was elevated to about 250°C.

3.2. Target for material irradiation facility

Material irradiation facility also requires intensive neutron source to test material tolerance of radiation damage. The China material irradiation facility aims at a small-scale and low-cost facility. To maximize the radiation effect, the facility design took the advantage of forward neutrons. The system adopted deuterium beam of 20 MeV@5 mA and beryllium as the target material. Due to 20 MeV deuterium only has a very low penetration at the order of millimeter; there is hardly effective neutrons leak out, with the previous configuration. Hence, a new structure based on chute flow was adopted.

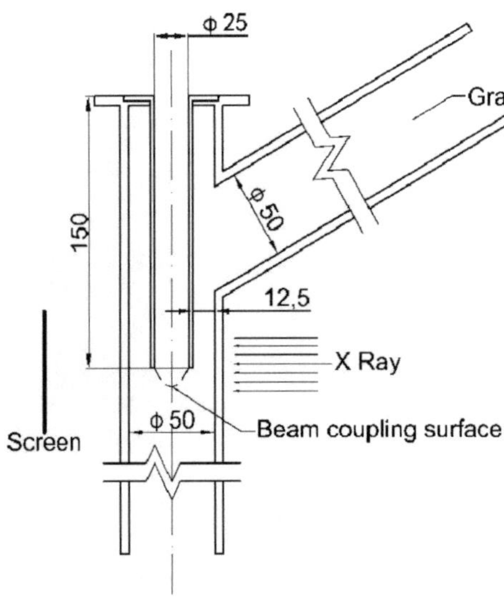

Figure 3. Profile of coupling section.

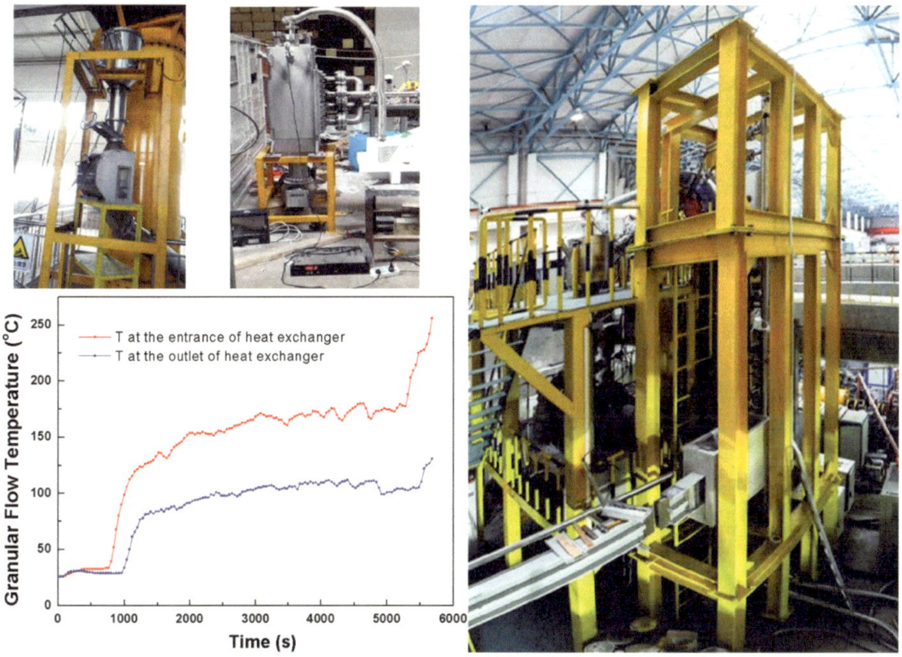

Figure 4. Prototype facility and heating test.

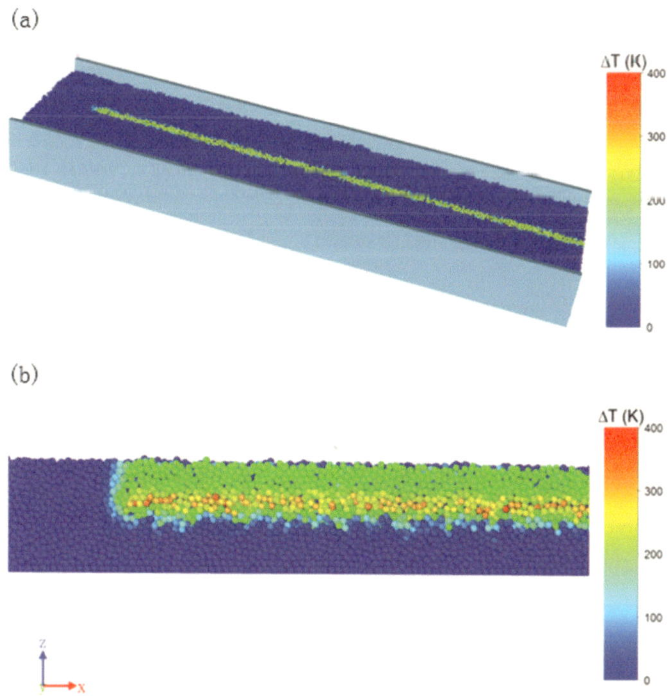

Figure 5. Simulation of beam heat load.

To demonstrate the heat load effect of this chute flow target, this progress is simulated by our GPU code [58]. **Figure 5** showed one of the simulation results of heat deposition in a chute flow. The result showed the 20 MeV beam particles only penetrate several millimeters into the flow. According to the design and simulation, mass flow at about 2–3 m/s is sufficient to remove the heat. The parameters of feeding rate, chute slope, beam spot position, and flow channel width system could be well adjusted to required reaction region. After its first test, target section of the facility in **Figure 2** was replaced by a slope chute. **Figure 6** showed the test loop facility and targets applied in the system. An acrylic chute was chosen to observe the flow status, and a stainless steel chute was installed for beam coupling effect of the facility. The experimental result is showed lower right.

Figure 6. Electron beam coupling of chute flow target.

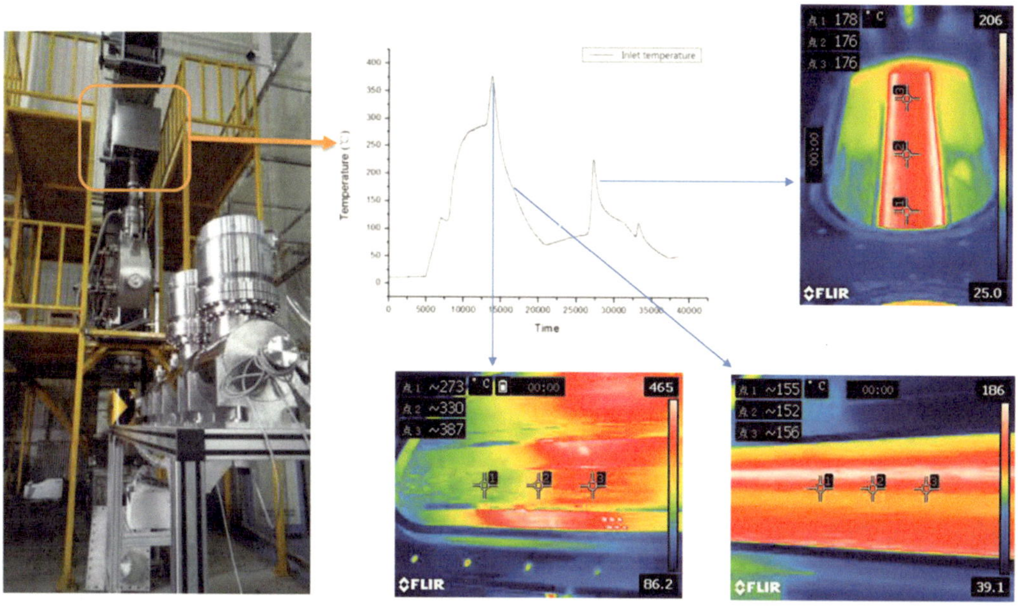

Figure 7. Prototype of material irradiation facility target and heating.

Figure 8. Static granular target.

The target design is first tested on remodeled small-scale circulation loop for ADS target in 2015. The facility is also radiated by electron beam. Temperature of the flow increased slightly, which showed that the system heat removal is workable.

Later in 2017, a full-scale prototype facility finished its major construction, as shown in **Figure 7**. Due to the radiation hazards, instead of actual beam, system used laser and electric heating to test temperature endurance and heat removal effect. 30 kW electric heating offered major overall heating power, while laser provided a high power density up to 3 MW/cm^2. Granular flow reached an average temperature more than 300°C and maximum temperature more than 500°C. High-temperature testing was performed over 200 h, among which the longest continuous run lasts more than 100 h.

To test the neutron performance of this target material, a static granular target as shown in **Figure 8** was installed in the beamline of a 20 MeV linear accelerator. The neutron yield data is tested through proton beam and the same 1 mm beryllium in static vacuum chamber. Five stacks of activated foils are located around the chamber. The measurement of the foils deduced a total neutron flux of $2.14\times10^{13}/cm^2\cdot mA\cdot s$.

4. Conclusion

The unique property of dense granular flow makes it possible to transfer heat effectively at high temperature. The major mechanism and factors of granular were reviewed. Using the granular material as heat transfer media and target material, two different configurations came up depending on the application. The prototype testing facilities were constructed. More designs and applications based on the granular heat transfer media are to be developed hopefully.

Acknowledgements

The major work mentioned above is achieved by the staff members from spallation target department and advance nuclear material department of Institute of Modern Physics of the CAS.

Notation

k	effective thermal conductivity
k_g	gas phase thermal conductivity
k_s	solid phase thermal conductivity
D_p	particle diameter
ε	void fraction/porosity

Author details

Xuezhi Zhang and Lei Yang*

*Address all correspondence to: lyang@impcas.ac.cn

Institute of Modern Physics of the Chinese Academy of Sciences, Lanzhou, China

References

[1] Bauer G. Overview on spallation target design concepts and related materials issues. Journal of Nuclear Materials. 2010;**398**(1-3):19-27

[2] Buligins L, Thomsen K, Lielausis O, et al. Internal geometry and coolant choices for solid high power neutron spallation targets. Nuclear Instruments & Methods In Physics Research Section A-Accelerators Spectrometers Detectors And Associated Equipment. 2014;**761**:58-68

[3] Hao J, Chen Q, Xu Y, et al. Target thickness optimization design of a spallation neutron source target cooling system. Applied Thermal Engineering. 2013;**61**(2):641-648

[4] Bauer G_o OECD. Towards reliable high power spallation target design. 2003. pp. 81-93

[5] Yang L, Zhan W. New concept for ADS spallation target: Gravity-driven dense granular flow target. SCIENCE CHINA Technological Sciences. 2015;**58**(10):1705-1711

[6] Incropera F. Fundamentals of Heat and Mass Transfer. John Wiley & Sons; 2011

[7] Lehr J. Introduction. In: Nuclear Energy Encyclopedia. Hobeken, NJ, USA: John Wiley & Sons, Inc; 2011. pp. xi-xii

[8] Wolf J. Supercritical water reactor. In: Nuclear Energy Encyclopedia. Hobeken, NJ, USA : John Wiley & Sons, Inc.; 2011. pp. 305-308

[9] Suzuki M, Kawasaki S. Development of computer code PRECIP-II for calculation of Zr-steam reaction: Comparison of calculation with experiments in temperature transient. Journal of Nuclear Science and Technology. 1980;**17**(4):291-300

[10] Marques JG. Safety of nuclear fission reactors: Learning from accidents. In: Nuclear Energy Encyclopedia. Hobeken, NJ, USA: John Wiley & Sons, Inc.; 2011. pp. 127-149

[11] Gromov B et al. Use of lead-bismuth coolant in nuclear reactors and accelerator-driven systems. Nuclear Engineering and Design. 1997;**173**(1-3):207-217

[12] Hill R, Grandy C, Khalil H. Generation-IV sodium-cooled fast reactors (SFR). In: Nuclear Energy Encyclopedia. John Wiley & Sons, Inc.; 2011. pp. 353-364

[13] Martone M. IFMIF: International Fusion Materials Irradiation Facility Conceptual Design Activity: Final report. ENEA, Frascati, Italy: Dipt. Energia; 1997

[14] De Bruyn D et al. From Myrrha to XT-ADS: The design evolution of an experimental ads system. AccApp'07; 2007. pp. 848-854

[15] Taylor P. Generation-IV gas-cooled fast reactor. In: Nuclear Energy Encyclopedia. John Wiley & Sons, Inc.; 2011. pp. 349-351

[16] Faraday M. On a peculiar class of acoustical figures; and on certain forms assumed by groups of particles upon vibrating elastic surfaces. Philosophical Transactions of the Royal Society of London. 1831;**121**:299-340

[17] Duran J. Sands, Powders, and Grains—An Introduction to the Physics of Granular Materials. 2000

[18] Davies T et al. The production and anatomy of a tungsten powder jet. Powder Technology. 2010;**201**(3):296-300

[19] Magistris M. Radiological considerations on multi-MW targets. Part II: After-heat and temperature distribution in packed tantalum spheres. Nuclear Instruments & Methods in Physics Research Section A—Accelerators Spectrometers Detectors and Associated Equipment. 2005;**545**(3):823-829

[20] Colominas S et al. Octalithium plumbate as breeding blanket ceramic: Neutronic performances, synthesis and partial characterization. Fusion Engineering and Design. 2012;**87**(5-6):482-485

[21] Donne M, Sordon G. Heat-transfer in pebble beds for fusion blankets. Fusion Technology. 1990;**17**(4):597-635

[22] Rousseau P, du Toit C, Landman W. Validation of a transient thermal-fluid systems CFD model for a packed bed high temperature gas-cooled nuclear reactor. Nuclear Engineering and Design. 2006;**236**(5-6):555-564

[23] Reimann J et al. Measurements of the Thermal Conductivity of Compressed Beryllium Pebble Beds. 2005

[24] Herbst T. Shear phenomena in granular random packings. Soil Science. 1964;**98**(4):280

[25] Wieghardt K. Experiments in granular flow. Annual Review of Fluid Mechanics. 1975; **7**:89-114

[26] Beverloo W, Leniger H, Vandevelde J. The flow of granular solids through orifices. Chemical Engineering Science. 1961;**15**(3-4):260

[27] Aldin N, Gunn D. The flow of non-cohesive solids through orifices. Chemical Engineering Science. 1984;**39**(1):121-127

[28] Nedderman R. Statics and Kinematics of Granular Materials. Cambridge University Press; 2005

[29] Rao K, Nott P, Sundaresan S. An Introduction to Granular Flow. New York: Cambridge University Press; 2008

[30] Gutiérrez G et al. Silo collapse: An experimental study. In: Traffic and Granular Flow'07. Springer; 2009. pp. 517-523

[31] Hall J, Richart F. Dissipation of Elastic Wave Energy in Granular Soils. 1963

[32] Broughton J, Kubie J. Heat-transfer mechanism as applied to flowing granular media. International Journal of Heat and Mass Transfer. 1976;**19**(2):232-233

[33] Balakrishnan A, Pei D. Heat transfer in gas-solid packed bed systems. 1. A critical review. Industrial & Engineering Chemistry Process Design and Development. 1979;**18**(1):30-40

[34] Schlunder E. Heat transfer to packed and stirred beds from the surface of immersed bodies. Chemical Engineering and Processing. 1983;**18**:31-53s

[35] Sullivan W, Sabersky R. Heat transfer to flowing granular media. International Journal of Heat and Mass Transfer. 1975;**18**(1):97-107

[36] Spelt J, Brennen C, Sabersky R. Heat transfer to flowing granular material. International Journal of Heat and Mass Transfer. 1982;**25**(6):791-796

[37] Yagi S, Kunii D. Studies on effective thermal conductivities in packed beds. AICHE Journal. 1957;**3**(3):373-381

[38] Ernst R. Wärmeübergang an Wärmeaustauschern im Moving Bed. Chemie Ingenieur Technik. 1960;**32**(1):17-22

[39] Schlunder E. Wärmeübergang an bewegte Kugelschüttungen bei kurzfristigem Kontakt. Chemie Ingenieur Technik. 1971;**43**(11):651-654

[40] Wunschmann J, Schlunder E. Heat transfer from heated plates to stagnant and agitated beds of spherical shaped granules under normal pressure and vacuum. In: Proc. 5th Int. Heat Transfer Conf.; Tokyo. 1974

[41] Martin H. Low Peclet number particle-to-fluid heat and mass transfer in packed beds. Chemical Engineering Science. 1978;**33**(7):913-919

[42] Du Toit C. Radial variation in porosity in annular packed beds. Nuclear Engineering and Design. 2008;**238**(11):3073-3079

[43] de Klerk A. Voidage variation in packed beds at small column to particle diameter ratio. AIChE Journal. 2003;**49**(8):2022-2029

[44] Gnielinski V. Heat and mass-transfer in packed-beds. Chemie Ingenieur Technik. 1980;**52**(3):228-236

[45] Sissom L, Jackson T. Heat exchange in fluid-dense particle moving beds. Journal of Heat Transfer. 1967;**89**(1):1

[46] Chukhano Z. Heat and mass transfer between gas and granular material. 2. International Journal of Heat and Mass Transfer. 1970;**13**(12):1805

[47] Aberdeen J, Laby T. Conduction of heat through powders and its dependence on the pressure and conductivity of the gaseous phase. Proceedings of the Royal Society of London Series a-Containing Papers of a Mathematical and Physical Character. 1926;**113**:459-447

[48] Kannuluik W, Martin L. Conduction of heat in powders. Proceedings of the Royal Society of London Series A-Containing Papers of a Mathematical and Physical Character. 1933;**141**(843):144-158

[49] Sokolov V, Yablokova M. Thermal-conductivity of a stationary granular bed with upward gas-liquid flow. Journal of Applied Chemistry of the USSR. 1983;**56**(3):551-553

[50] Gelperin J, Kagan A. Study of heat transfer in gas flow in tubes packed with granular materials. Chemical Engineering Progress. 1966;**62**(8):88

[51] Zumbrunnen D, Viskanta R, Incropera F. Heat-transfer through granular beds at high-temperature. Warme Und Stoffubertragung-Thermo and Fluid Dynamics. 1984;**18**(4):221-226

[52] Jaguaribe E, Beasley D. Modeling of the effective thermal conductivity and diffusivity of a packed bed with stagnant fluid. International Journal of Heat and Mass Transfer. 1984;**27**(3):399-407

[53] Einav I et al. 10,000—A reason to study granular heat convection. In: Yu A et al., editors. Powders and Grains 2013; 2013. pp. 38-45

[54] Botterill J et al. The effect of gas and solids thermal properties on the rate of heat transfer to gas-fluidized beds. In: Proc. Int. Symp. on Fluidization; Eindhoven. Amsterdam: Netherlands University Press; 1967

[55] Baird M et al. Heat transfer to a moving packed bed of nickel pellets. Canadian Journal of Chemical Engineering. 2008;**86**(2):142-150

[56] Tian Y, Qi J, Lai J, et al. A heterogeneous CPU-GPU implementation for discrete elements simulation with multiple GPUs. In: 2013 International Joint Conference on Awareness Science and Technology and Ubi-Media Computing (i CAST-UMEDIA); 2013. pp. 547-552

[57] Roberts I. Determination of the vertical and lateral pressures of granular substances. Proceedings of the Royal Society of London. 1883;**36**(228-231):225-240

[58] Yagi S, Kunii D, Wakao N. Studies on axial effective thermal conductivities in packed beds. AIChE Journal. 1960;**6**(4):543-546

Chapter 5

Bubble Dynamics in a Narrow Rectangular Channel

Xu Jianjun, Xie Tianzhou, Chen Bingde and Bao Wei

Additional information is available at the end of the chapter

http://dx.doi.org/10.5772/intechopen.74608

Abstract

It is very important to study the bubble dynamics in order to understand the physical process of boiling heat transfer in a narrow channel. Experimental and theoretical studies on bubble dynamics in a narrow rectangular channel are proposed. The cross section of the narrow rectangular channel is 2 mm × 8 mm. A high speed digital camera is applied to capture bubble behaviors from the narrow side and wide side of the narrow rectangular channel. Bubble growth rate, bubble departure diameter, bubble interface parameter and others are obtained according to the observation. A force balance analysis on a growing bubble is proposed to predict the bubble departure diameter and sliding bubble velocity, and the predicted results agree with the experimental data. Thus, the mechanism of bubble departure, slide and lift-off behavior in a narrow rectangular channel can be explained by the analysis of forces.

Keywords: narrow channel, bubble dynamics, force balance

1. Introduction

Thermal hydraulics and safety analysis are important in nuclear reactor design and operation. There have been active investigations on nuclear thermal hydraulics and safety analysis in different core fuel elements. The rod bundles with grid spacers and plate-type fuel elements are frequently encountered at the aspect of developing new type nuclear reactor. The coolant channels of the plate-type fuel element are composed of some typical narrow rectangular channels, and the flow and heat transfer in a narrow rectangular channel is also widely employed in the compact evaporators and heat exchangers. So it is important to perform the thermal hydraulic study of boiling heat transfer in the narrow rectangular channels.

During the last decades, there were a lot of investigations focusing on two-phase flow and boiling heat transfer in narrow rectangular channels in published literatures [1–14]. The results show that two-phase flow and boiling heat transfer in a narrow channel is a very complex physical process. Some researchers think that boiling heat transfer in a narrow channel are different from that in an ordinary sized channel and the boiling heat transfer mechanism in a narrow rectangular channel is not fully understood.

In order to understand the physical process of boiling heat transfer in a narrow channel, and it is important to study the bubble growth, departure and lift-off behavior by using the visualized technology. Over the last half century, a lot of studies of bubble dynamics have been conducted, and this kind of study is significant to understand the boiling heat transfer mechanism. It is the fundamental for understanding and predicting the boiling heat transfer correctly. However, most of these studies have been focused on observation of bubble behaviors in an ordinary sized channel. In our previous work, bubble characteristics in narrow rectangular channel were visually studied, and the results showed that the bubbles departing from the nucleation sites always slid along the heating wall [15]. Some researchers presented the importance of the sliding bubble to enhance heat transfer in nucleate boiling [16–20], the results showed that turbulence caused by sliding bubbles could be an important mechanism in enhancing heat transfer. The effect of sliding bubbles on boiling heat transfer in a narrow channel will be more important than that in an ordinary sized channel because of the smaller cross section size.

It is known that the analysis of forces acting on a single bubble is a promising approach to sufficiently understand the mechanism of bubble motion. Levy [21] first proposed a model for the prediction of bubble departure diameter based on the balance of forces, including the buoyancy, surface tension force and drag force. Klausner et al. [22] built the models of force balance to predict the bubble departure diameter and lift-off diameter, and there have been active investigations on the analysis of forces balance [14, 23–25], and these researches implies that bubble interface parameters were required as the input parameters in order to solve this model, including the bubble inclination angle, upstream contact angle, downstream contact angle and bubble contact diameter. It is also important to develop the wall heat flux partitioning model and the interfacial area transport equation for using computational fluid dynamics (CFD) code.

These studies in the above literatures have not provided an adequately detailed understanding of boiling heat transfer in a narrow rectangular channel. The present study showed bubble growth, departure, sliding and lift-off behaviors in a narrow rectangular channel, and the mechanism of bubble departure and lift-off from the heating wall was discussed based on the analysis of forces acting on the bubble.

2. Experimental facilities and measuring system

Figure 1 shows a schematic of the test section of the narrow rectangular channel. The cross section of the narrow rectangular channel was $2 \times 8\ mm^2$, which was fabricated in an optical quartz glass. Both the wide side and the gap side of the optical quartz glass were polished to improve the transparency. Bubble behavior was visually observed from the wide side and gap

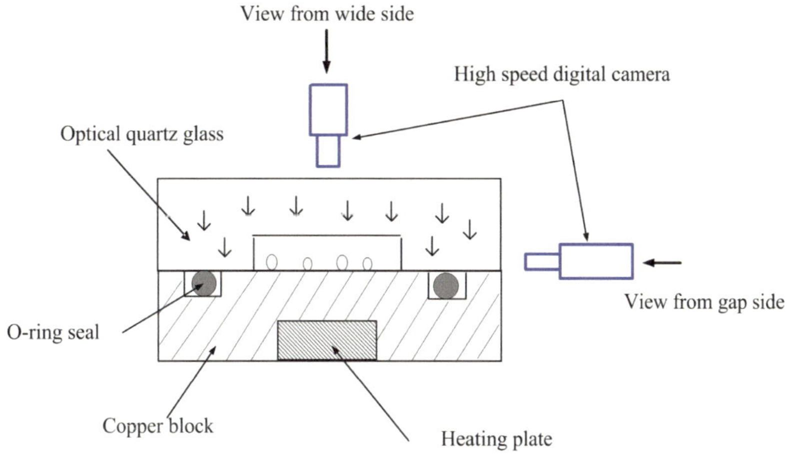

Figure 1. Schematic diagram of test section.

side of the rectangular channel. The observation windows were made of optical quartz glass. The width of the narrow rectangular channel is 8 mm and height is 2 mm. The same experimental apparatus was described to investigate the bubble dynamics in the previous study of the authors [13].

In the present work, the experiment pressure was atmospheric pressure. The range of the liquid subcooling of the inlet of the test section was 10–40°C. The range of mass flux was 129–706 kg/m^2 s. Temperatures at the wall, inlet, and outlet of the test section were registered by 1 mm diameter T-type thermocouples, which were calibrated and a uncertainty of 0.5°C for all thermocouples. The uncertainty in the heat supplied to the test section was less than ±0.9%. The heat loss of the test section was estimated to within 7% of the total of the heating power in the present study. Flow rate was measured by a venturimeter and the uncertainty was less than

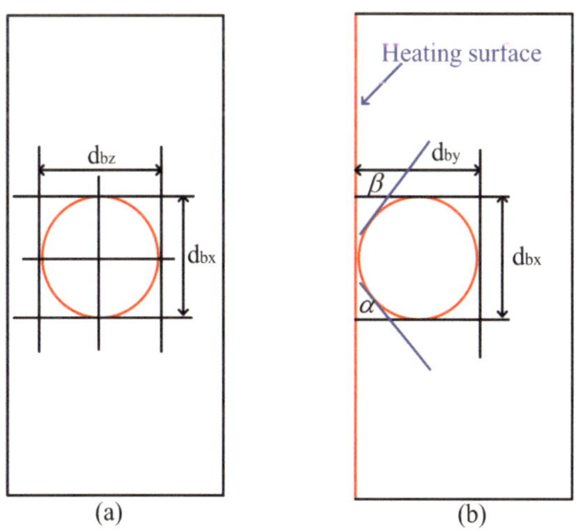

Figure 2. Definition of bubble equivalent diameter.

±0.8% of the full scale range. The pressure of test section was measured by a pressure transmitter, and the uncertainty was less than ±0.5 kPa of the full scale range.

A high speed digital camera (Micro-NIKKOR lens 200 mm 1:1) was used to capture bubble dynamics. The successive typical bubble images were obtained by using a high speed digital camera. The bubble dynamics can be obtained from the wide side and gap side of the narrow rectangular channel, respectively. A picture of the vernier caliper was captured with the high speed digital camera, and the scale factor was defined by the image of vernier caliper. The real dimension of bubble was obtained with the scale factor. The bubble dynamics was captured with a speed of 3800 fps (frames per second). The 576 × 576 pixel picture corresponds to a 7.2 mm × 7.2 mm recorded field in real dimension. **Figure 2** defined the bubble equivalent diameter as following:

$$\frac{\pi}{6}d^3 = \frac{\pi}{6}d_{bx}d_{bz}^2 \Rightarrow d = \sqrt[3]{d_{bx}d_{bz}^2} \quad (if \quad d_{bx} > d_{bz}) \tag{1}$$

$$\frac{\pi}{6}d^3 = \frac{\pi}{6}d_{bx}d_{by}^2 \Rightarrow d = \sqrt[3]{d_{bx}d_{by}^2} \quad (if \quad d_{bx} > d_{by}) \tag{2}$$

The ratio R of the bubble contact diameter to the bubble departure diameter is expressed as:

$$R = d_w/d_0 \tag{3}$$

3. Experimental observation of bubble dynamics

3.1. Bubble growth and departure

A series of images of a typical bubble growing at the nucleation site is obtained by observation from the narrow side of the rectangular channel, as shown in **Figure 3**. The bubble shape is almost spherical when the bubble is growing at the nucleation sites. The contact diameter between the bubble base and heating wall is observed when the bubble is growing at the nucleation site. The interface parameters of typical bubbles just departing from the nucleation

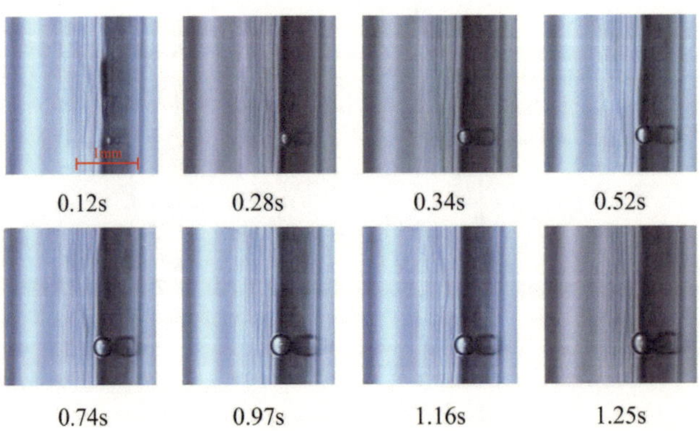

Figure 3. Visualization of bubble growth and departure from narrow side.

http://dx.doi.org/10.5772/intechopen.74608

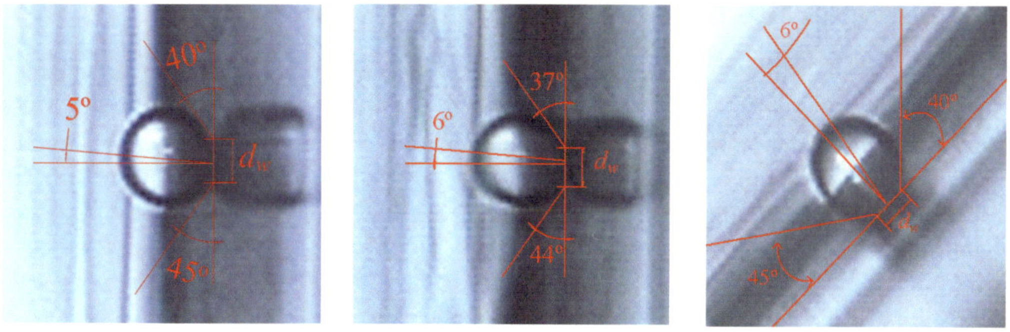

(a)Vertical upflow ,d=0.22mm (b) Vertical upflow,d=0.176mm (c) Inclined upward facing upflow,d=0.145mm

Figure 4. Bubble contact angle and inclination angle [14].

site, such as the bubble inclination angle, upstream contact angle, downstream contact angle, and bubble contact diameter, are measured by observation [14], as shown in **Figure 4**. The bubble upstream contact angle varies from 44° to 45°, the bubble downstream contact angle varies from 37° to 40°, the inclination angle varies from 5° to 6°, the range of R in the current experiment is about 0.41–0.5, and the average of R is about 0.45.

Figure 5 shows the effect of heat flux on bubble growth at the same nucleation site [14]. The bubble growth rate increases with increasing heat flux. The bubble departure time decreases with increasing heat flux, and so the bubble departure diameter decreases with increasing heat flux. According to Zuber's model [26], the bubble growth rates can be approximated by a power law curve fit, and the range of the power law varies from about 0.34 to 0.684.

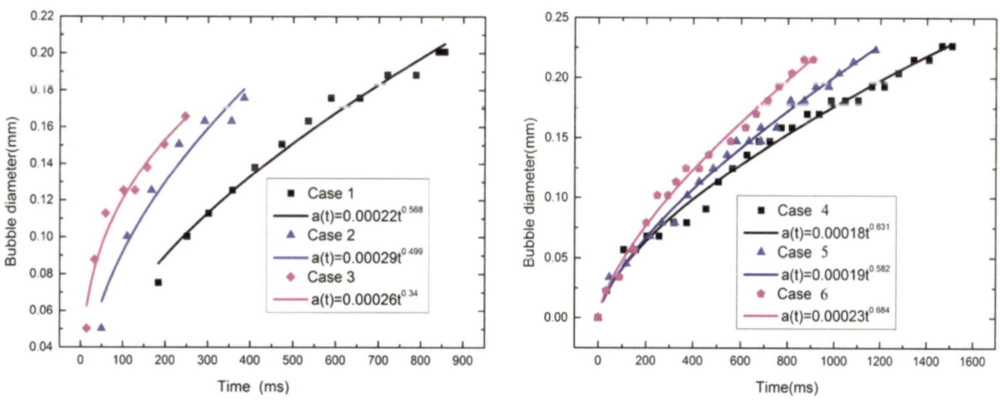

(a) Effect of heat flux on bubble growth (b) Effect of heat flux on bubble growth

Case1: $G=140.3kg/m^2 \cdot s, T_{in}=77.3℃, q=26.3kW/m^2$ Case2: $G=139.2kg/m^2 \cdot s, T_{in}=79.5℃, q=31.1kW/m^2$
Case3: $G=137.7kg/m^2 \cdot s, T_{in}=78.7℃, q=40.4kW/m^2$ Case4: $G=253.6kg/m^2 \cdot s, T_{in}=74.5℃, q=46.3kW/m^2$
Case5: $G=252.8kg/m^2 \cdot s, T_{in}=74.4℃, q=46.7kW/m^2$ Case6: $G=250.0kg/m^2 \cdot s, T_{in}=73.6℃, q=56.5kW/m^2$

Figure 5. Bubble growth diameter with time [14].

3.2. Sliding bubble behavior in vertical channel

The images of typical sliding bubbles motion at the same observation window are obtained from the narrow side of the rectangular channel, as shown in **Figure 6**. After a bubble departs from the nucleation site, it always slides along the heating wall. The typical bubble sliding phenomenon in vertical upflow boiling is observed at low heat flux in the isolated bubble region [13]. During the bubble sliding along the heating wall, the bubble contact diameter between the bubble base and heating wall is observed. The sliding bubble diameter increases with increasing heat flux. The number of sliding bubbles increases gradually with increasing heat flux. The sliding bubbles start to coalescence, and sliding bubbles become the larger after coalescence, kenning et al. [19] considered it as the main model for influence of the thermal boundary layer reformation near the wall due to sliding bubble motion. The phenomenon of bubble lift-off from the heating wall is not observed, which shows the effect of the sliding bubble interaction among the sliding bubbles is also little in the isolated bubble region with low heat flux. It is known that the turbulent effect is strong with increasing number of bubbles for the high heat flux, which maybe results in bubble lift-off from the heating wall.

The shape of the small sliding bubble appears spherical, but it is elongated in the direction normal to the heating wall with increasing bubble diameter. When the bubble slides along the heating wall, the upstream contact angle and downstream contact angle of the sliding bubble is almost equal, and so the inclination angle of the sliding bubble approaches to zero. In the present work, the confined bubble growing and sliding motion in the narrow rectangular channel is not observed.

Figure 7 shows the sliding bubble velocity under different bubble diameter with time, as is seen from **Figure 7**, the sliding bubble velocity increase obviously at the initial moment, but the

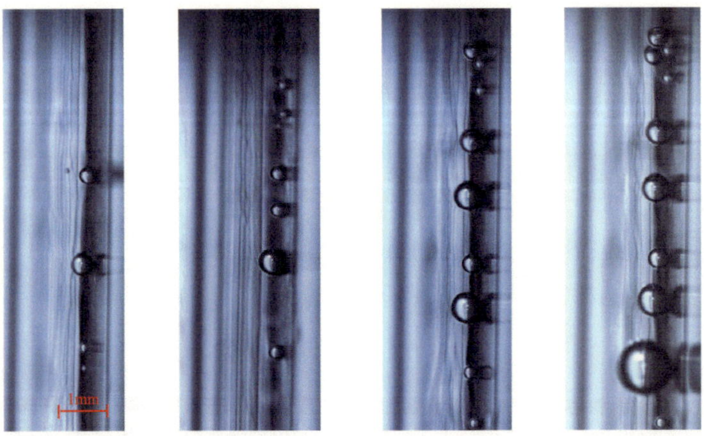

(a)T_{in}=80.4℃, q=33.7kW/m^2 (b)T_{in}=78.9℃, q=42kW/m^2 (c)T_{in}=78.1℃, q=52kW/m^2 (d)T_{in}=77℃, q=58.6kW/m^2

G≈139.6 kg/(m^2·s)

Figure 6. Typical sliding bubble with increasing heat flux.

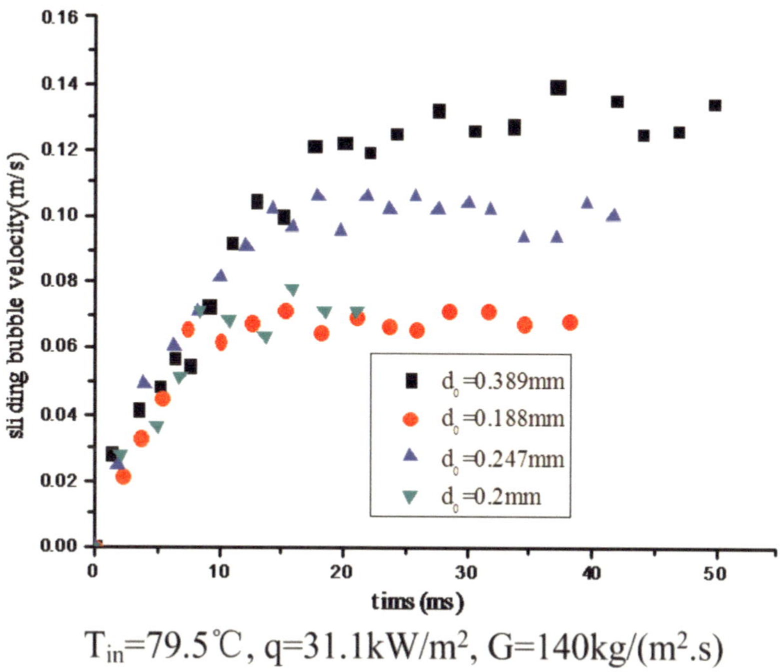

Figure 7. Sliding bubble velocity with time.

sliding bubble velocity approach asymptote, ultimately. The influence of the sliding bubble diameter on the velocity is larger. The sliding bubble ultimate velocity increase with increasing bubble diameter, which shows the importance of buoyancy in the flow direction. The velocity of sliding bubble ($d_0 = 0.389$ mm) approaches to the bulk liquid velocity.

3.3. Bubble lift-off behavior

In inclined 45° upward facing upflow boiling, the phenomenon on bubble lift-off from the heating wall is observed from the narrow side of the rectangular channel at the same observation window, as is shown in **Figure 8**. After a bubble slides some distance, it tends to lift-off from the heating wall, and the buoyancy at the direction normal to the surface is responsible for the bubble lift-off in inclined 45° upward facing heating wall. The bubble velocity will increase when the bubble lifts off from the heating wall. The condensation phenomenon on the interface of the lift-off bubble occurs when the bulk flow is subcooled, which results in the change of bubble shape.

In vertical downflow boiling, the images of bubbles are obtained from the same observation window with the different heat flux, as is shown in **Figure 9**. After a bubble departs from the nucleation site, it tends to lift-off from the surface. When the bubble lifts off from the heating wall, the shadow of the bubble is observed on the heating wall. The bubble size is small due to the condensation on the interface of the lift-off bubble. The bubble velocity in vertical downflow boiling is small due to the opposite direction of the buoyancy and drag forces acting on the

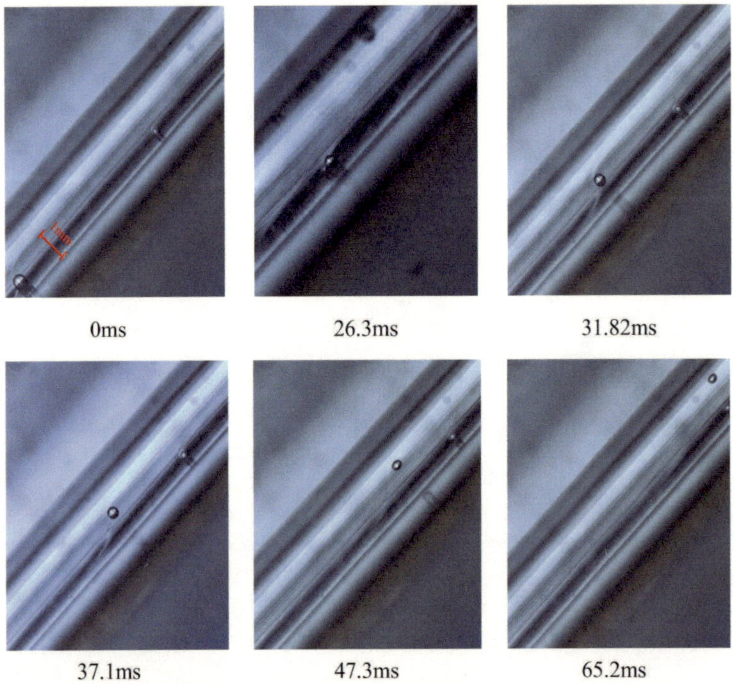

Figure 8. Single bubble motion in inclined 45° upward facing upflow boiling.

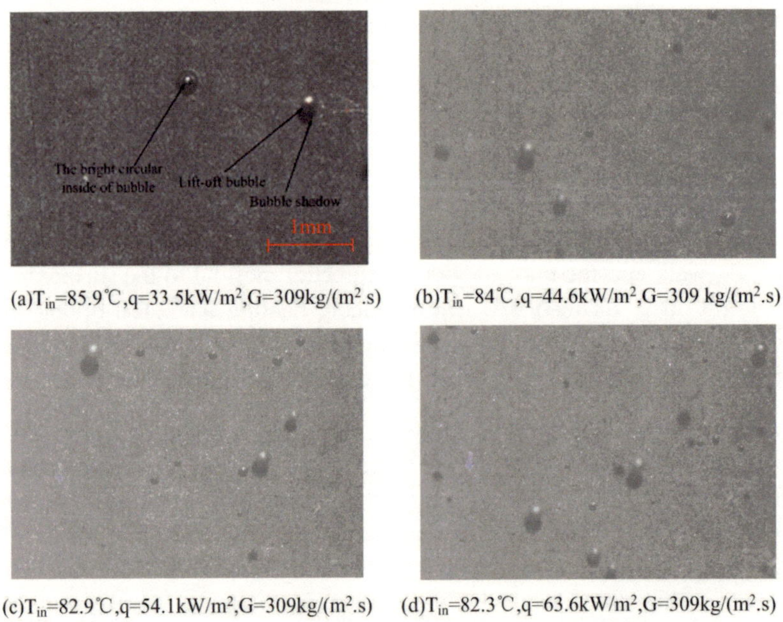

Figure 9. Bubble motion in vertical downflow boiling.

bubble. The local fluid velocity is always higher than that of the bubble velocity, and so the shear lift force is the driving force to promote bubble lift-off from the heating wall.

4. Analysis of force acting on the bubble

4.1. Balance of forces acting on a single bubble

The forces acting on a single bubble are schematically shown in **Figure 10**. Different forces acting on a bubble in the directions parallel and normal to a heating wall are analyzed, and the force in the x and y directions are evaluated as following:

$$\sum F_x = F_{bx} + F_{sx} + F_{amx} + F_{qs} = m_b \frac{dv_x}{dt} \tag{4}$$

$$\sum F_y = F_{bx} + F_{sy} + F_{amy} + F_{sl} + F_h + F_{cp} = m_b \frac{dv_y}{dt} \tag{5}$$

where F_b is the buoyancy force, F_s is the surface tension force, F_{am} is the added-mass force, F_{qs} is the quasi-steady drag force, m_b is the mass of the sliding bubble, v_x is the velocity of displacement of the center of mass of the bubble in the x direction, t is the time, F_{sl} is the shear lift force, F_h is the force due to the hydrodynamic pressure, F_{cp} is the contact pressure force, θ is the incline angle of the bubble, α is the upstream contact angle, β is the downstream contact angle, v_y is the velocity of displacement of the center of mass of the bubble in the y direction, d_w is the contact diameter, $u(y)$ is the local liquid velocity profile near the wall, and Φ is the incline angle of the heating wall.

4.1.1. *Buoyancy*

The buoyancy acting on a bubble due to the surrounding liquid is expressed as

$$F_b = (\rho_l - \rho_v)V_b g \tag{6}$$

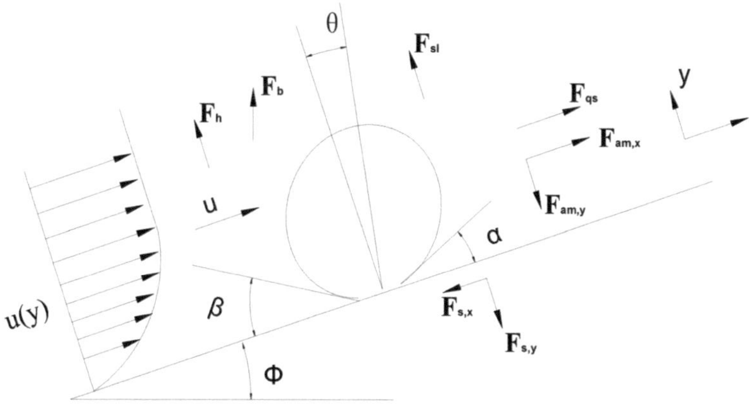

Figure 10. Force acting on a bubble.

where ρ_l is the density of liquid, ρ_v is the density of the vapor bubble, g is the gravitational acceleration, and V_b is the bubble volume.

4.1.2. Surface tension force

The surface tension forces at x and y directions are given as follows [22]:

$$F_{sx} = -d_w \sigma_{lv} \frac{\pi}{\alpha - \beta}(\cos\beta - \cos\alpha) \tag{7}$$

$$F_{sy} = -1.25 d_w \sigma_{lv} \frac{\pi(\alpha-\beta)}{\pi^2 - (\alpha-\beta)^2}(\sin\alpha + \sin\beta) \tag{8}$$

where σ_{lv} is surface tension.

4.1.3. Added-mass force

In order to estimate the added-mass force, Thorncroft and Klausner [23] developed the following expression by solving the inviscid flow problem for a growing sphere in a uniform, unsteady flow.

$$F_{am} = \frac{1}{2}\cdot\frac{4}{3}\pi\rho_l a^3\left(\frac{dU}{dt} - \frac{dv_x}{dt}\right) + 2\pi\rho_l a^2 (U - v_x)\dot{a} \tag{9}$$

where the first term is the conventional added-mass force. The second term is the added-mass force due to bubble expansion. In the steady-state flow, dU/dt approaches zero. U is bulk liquid velocity, v_x is bubble velocity and a is the bubble growth radius.

When the bubble is growing at the nucleation site, the bubble's center of mass moves away from the heating wall and the added-mass force leads to the unsteady force in the direction normal to the wall. In keeping with convention, the added-mass force is referred to as the bubble growth force. Zeng et al. [27] suggested the growth force can be estimated from:

$$F_{am} = -\rho_l \pi a^2\left(\frac{3}{2}C_S\dot{a}^2 + a\ddot{a}\right) \tag{10}$$

where $C_S = 20/3$.

4.1.4. Quasi-steady drag force

Delnoij et al. [28] suggested that the quasi-steady drag force can be estimated from:

$$F_{qs} = \frac{1}{2}C_D\rho_l\pi a^2(u - v_x)|u - v_x| \tag{11}$$

$$C_D = \begin{cases} 240 & \mathrm{Re}_b \le 0.1 \\ \frac{24}{\mathrm{Re}_b}\left(1 + 0.15\mathrm{Re}_b^{0.687}\right) & 0.1 < \mathrm{Re}_b \le 1000 \\ 0.44 & \mathrm{Re}_b > 1000 \end{cases} \tag{12}$$

$$\mathrm{Re}_b = \frac{2a(t)|u - v_x|}{\gamma} \tag{13}$$

where γ is the liquid kinematic viscosity, u is the velocity of local liquid, and Re_b is the bubble Reynolds number.

The local liquid velocity of the mass center of the bubble is estimated by using universal single-phase flow profile. The dimensionless velocity of different region is expressed by the following correlations:

$$u^+ = \begin{cases} y^+ & y^+ \le 5 \\ 5\ln y^+ - 3.05 & 5 < y^+ \le 30 \\ 2.5\ln y^+ + 5.5 & y^+ > 30 \end{cases} \tag{14}$$

$$u^+ = \frac{u}{u^*} = \frac{u}{\sqrt{\tau_w/\rho_l}} \tag{15}$$

$$y^+ = \frac{yu^*}{\gamma} = \frac{y\sqrt{\tau_w/\rho_l}}{\gamma} \tag{16}$$

$$u^* = \sqrt{\tau_w/\rho_l} \tag{17}$$

The wall shear stress can be calculated by the following expression

$$\tau_w = \frac{1}{2} C_f \rho_l U^2 \tag{18}$$

where C_f is the friction coefficient, which is calculated by the following expression

$$C_f = \lambda/4 \tag{19}$$

where λ is the friction factor. For a smooth surface, the friction factor is expressed by the following expression

$$\lambda = \begin{cases} 64/\mathrm{Re} & \mathrm{Re} < 2000 \\ 0.3163/\mathrm{Re}^{0.25} & 2000 \le \mathrm{Re} \le 100000 \end{cases} \tag{20}$$

$$\mathrm{Re} = \frac{UD_h}{\gamma} \tag{21}$$

where D_h *is* the hydraulic equivalent diameter.

4.1.5. Shear lift force

Saffman [29] developed an expression for the lift force on a particle in a low Reynolds number shear flow. Recently, Mei and Klausner [30] modified the Saffman model for a bubble as follows:

$$F_{sl} = \frac{1}{2} C_l \rho_l \pi a^2 (u - v_x)|u - v_x| \tag{22}$$

where C_l is the shear lift force coefficient and is defined as

$$C_l = 3.877 G_s^{1/2} \left(\mathrm{Re}_b^{-2} + 0.014 G_s^2\right)^{1/4} \tag{23}$$

where

$$G_s = \left|\frac{du}{dy}\right| \frac{a(t)}{u - v_x} = \frac{1}{u - v_x} \frac{1}{k^+ u^+} \tag{24}$$

4.1.6. Contact pressure force

The contact pressure force due to the pressure difference inside and outside the bubble at the reference point over the contact area is given as follows [22]:

$$F_{cp} = \frac{\pi d_w^2}{4} \frac{2\sigma_{lv}}{r} \tag{25}$$

4.1.7. Hydrodynamic pressure force

Klausner et al. [22] suggested that the hydrodynamic force can be estimated from

$$F_h = \frac{19}{24} \rho_l (u - v_x)|u - v_x| \frac{\pi d_w^2}{4} \tag{26}$$

The criteria of the bubble departing and sliding from the nucleation site was proposed by Klausner et al. [22]. The bubble remains attached to the nucleation site when the conditions $\sum F_x < 0$ and $\sum F_y < 0$. When the condition $\sum F_x = 0$ is violated prior to $\sum F_y = 0$, the bubbles will slide along the heating wall, and the criterion for bubble departure from the nucleation site is defined according to the condition $\sum F_x = 0$, at which point the bubble will departure from the nucleation site. When the condition $\sum F_y = 0$ is violated prior to $\sum F_x = 0$, the bubbles will lift-off from the heating wall without sliding, and the criterion for bubble lift-off from the heating wall is defined according to the condition $\sum F_y = 0$, at which point, the bubble will lift-off from the heating wall.

4.2. Analysis of mechanism of bubble departure

In order to predict the bubble departure diameter by analysis of force acting on the bubble, the interface parameters of a typical bubble, such as bubble contact diameter, d_w, bubble inclination angle, θ, upstream and downstream contact angle, α and β, must be measured. Based on the experimental observation of bubble behavior by Xu et al. [14], d_w, θ, α and β can be set as $0.45d_b$, 45°,40°, and 10°, respectively. The bubble departure diameters under different thermal hydraulic conditions are predicted by solving the balance model of forces. The predictions of the bubble departure diameter are compared by 48 experimental data in both vertical and inclined conditions. **Figure 11** shows the predicted result agrees with experimental data.

The forces acting on the bubble in the x direction and y direction are calculated by using the balance model of forces in order to analyze the mechanism of bubble departing from the nucleation sites. **Table 1** shows the value of different forces acting on the bubble just departing from the nucleation sites.

The prediction of results indicates that a bubble will slide along the heating wall before lift-off because the condition $\sum F_x = 0$ is violated prior to the condition $\sum F_y = 0$, which is accorded with the visual experimental observation. As is seen from **Table 1**, the bubble growth force is much less than that of other forces due to the lower growth rate of the bubble, and it is negligible in the current case. The buoyancy force, quasi-steady drag force, and surface tension force are the main forces controlling the bubble departure.

4.3. Analysis of mechanism of sliding bubble motion

According to the experimental results, the inclination angle of the bubble approached 0 during the bubble sliding along the heating wall, and so the x components of the growth and surface tension forces are approximately 0. The balance of forces acting on a single bubble at the x direction is expressed as

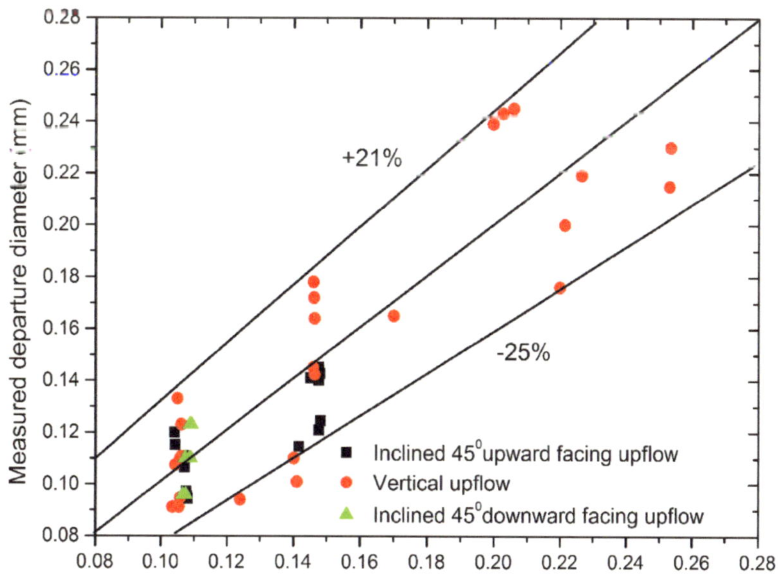

Figure 11. Comparison between predicted and measured results [14].

F_{bx}	F_{qs}	F_{sx}	F_{amx}	ΣF_x	F_{by}	F_{sy}	F_{amy}	F_{sl}	F_h	F_{cp}	ΣF_y
1.4E-08	6.1E-08	-7.5E-08	-1.1E-12	8.0E-11	1.1E-09	-8.5E-06	-9.9E-14	3.1E-08	3.2E-08	1.2E-06	-7.3E-06
1.4E-08	6.3E-08	-7.7E-08	-1.1E-12	1.3E-12	4.6E-09	-8.8E-06	-9.7E-14	3.2E-08	3.5E-08	1.2E-06	-7.5E-06
1.6E-08	7.3E-08	-8.9E-08	-5.6E-13	1.8E-10	4.5E-09	-1.0E-05	-4.9E-14	3.8E-08	5.2E-08	1.6E-06	-8.4E-06
2.0E-09	7.9E-08	-8.1E-08	-2.2E-12	2.1E-10	-2.6E-10	-4.6E-06	-2.0E-13	9.1E-08	6.1E-08	6.5E-07	-3.8E-06

Table 1. The force acting on the typical bubble (unit: N).

$$(\rho_l - \rho_v)V_b g \sin\Phi + \frac{1}{2}C_D\rho_l\pi a^2(u - v_x)|u - v_x| - \frac{1}{2}\cdot\frac{4}{3}\pi\rho_l a^3\frac{dv_x}{dt} + 2\pi\rho_l a^2(U - v_x)\dot{a} = m_b\frac{dv_x}{dt} \quad (27)$$

In order to solve the above differential equation, the initial condition is defined as $v(t_{depart}) = 0$, i.e., v is 0 when the vapor bubble just departs from the nucleation site and t_{depart} is the departure time of the vapor bubble. In our previous work [31], the sliding bubble velocities for pool and flow boiling in an ordinary size channel were obtained and the predicted results agreed with the experimental data of Maity [32]. However, the above balance of force is (27) was directly used to calculate the sliding bubble velocity in a narrow rectangular channel, it is found that the predicted result is not reasonable, and the force acting on the bubble is need to re-analyzed. The added-mass force is associated with the bubble growth rate and rate of change of bubble velocity. Some researchers showed that flow and heat transfer in a narrow rectangular channel were different with that in an ordinary size channel. In this experiment, the results show that the bubble grows slowly in a narrow rectangular channel and the bubble departure time is longer. This indicates that the size of a narrow channel inhibits bubble growth and departure. While the added-mass force were developed in an ordinary size channel by Thorncroft et al. [23]. Based on 72 experimental data on the sliding bubble velocity in a narrow rectangular channel, the value of the empirical constant C used for the evaluation of added-mass force is proposed, and the added-mass forces is expressed as:

$$F_{am} = C\left[\frac{1}{2}\cdot\frac{4}{3}\pi\rho_l a^3\left(\frac{dU}{dt} - \frac{dv_x}{dt}\right) + 2\pi\rho_l a^2(U - v_x)\dot{a}\right] \quad (28)$$

When the empirical constant C is 13.4, the prediction results agree with the experimental results.

Figure 12 shows the sliding bubble velocity against time for vertical flow boiling under the conditions of a bulk velocity of 0.143 m/s. As is seen from **Figure 12**, the predicted trends of the sliding bubble velocity agree with the experimental data. The sliding bubble velocity increases with increasing time, but the trend of an increase in the sliding bubble velocity decreases gradually as time increases. At about 14 ms, the sliding bubble velocity is higher than the local liquid velocity of the center of mass of the bubble. At about 50 ms, the sliding bubble velocity is almost equal to the bulk velocity. **Figure 13** shows the forces acting on the sliding bubble against time. The buoyancy gradually increases with increasing time, but the trend of an increase in the buoyancy is small, this is because the bubble growth rate is low. The buoyancy is the driving force to promote the bubble to slide along the surface in vertical flow boiling. At initial moment, the quasi-steady drag force is the driving force to promote the bubble to slide along the surface. At about 14 ms, the sliding bubble velocity exceeds the local liquid velocity

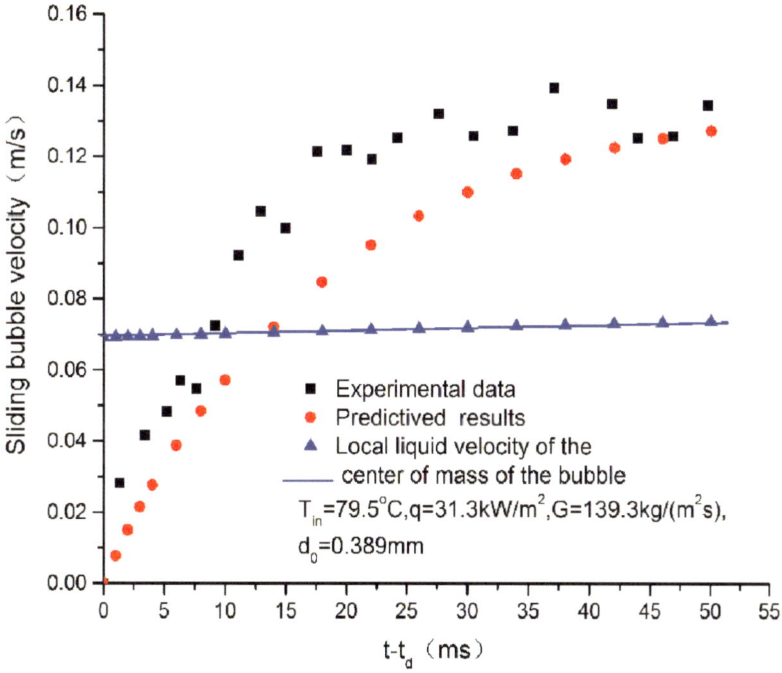

Figure 12. Sliding bubble velocity for vertical flow.

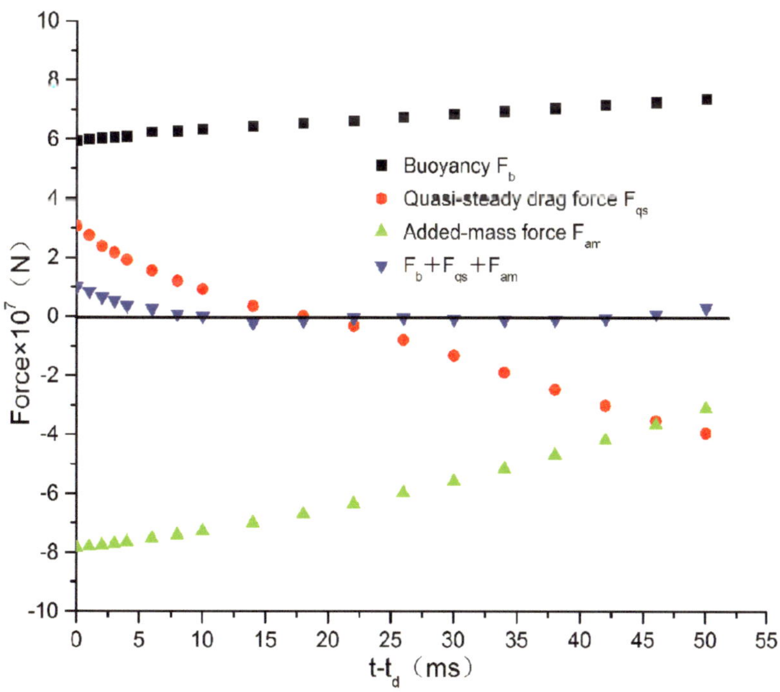

Figure 13. Forces acting on the sliding bubble for vertical flow.

and therefore, the quasi-steady drag force will become the resistance, which prevents the bubble from sliding along the surface. The quasi-steady drag force gradually increases because of an increasing difference between the sliding bubble velocity and the liquid velocity. During this process of the bubble sliding motion, the added-mass force is always a resistance to prevent the bubble from sliding along the surface. As the time increases, since the sliding bubble velocity exceeds the local liquid velocity, and therefore the shear lift force in the y direction will push the bubble against the wall, maybe this is cause of the bubble sliding along the surface in a vertical channel.

Figure 14 shows the sliding bubble velocity against time for in inclined 45° upward facing upflow boiling under the conditions of a bulk velocity of 0.143 m/s. As is seen from **Figure 14**, the predicted trends of the sliding bubble velocity agree with the experimental data. The trends of the sliding bubble velocity against time are similar to the above in **Figure 14**. The sliding bubble velocity increases quickly at the initial moment, but the trend of an increase in the sliding bubble velocity decreases gradually as the time increases. The buoyancy in the x direction is the driving force to promote the bubble to slide along the surface, as shown in **Figure 15**. At about 5 ms, the sliding bubble velocity is higher than the local liquid velocity of the center of mass of the bubble, and therefore the quasi-steady drag force will become the resistance, which prevents the bubble from sliding along the surface. At initial moment, the added-mass force is a resistance to prevent the bubble from sliding along the surface. At about 9 ms, the added-mass force will become a driving force to promote the bubble to slide along the surface. As the time increases, since the sliding bubble velocity exceeds the local liquid

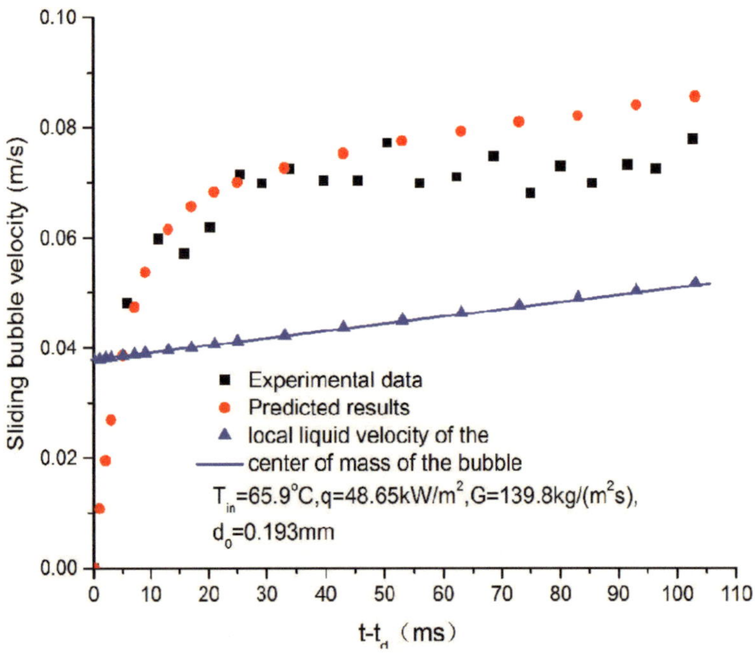

Figure 14. Sliding bubble velocity in inclined 45° upward facing upflow.

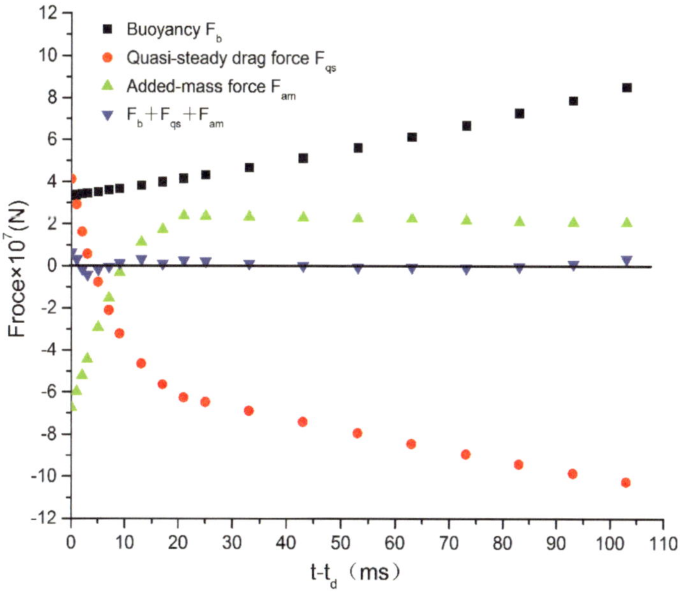

Figure 15. Force acting on the sliding bubble in inclined 45° upward facing upflow.

velocity of the center of mass of the bubble, and therefore the shear lift force in the y direction will push the bubble against the wall. The buoyancy in the y direction promotes the bubble lift-off from the surface.

5. Conclusions

Experimental and theoretical studies on bubble dynamics in a narrow rectangular channel are proposed in this chapter. A high speed digital camera is applied to capture bubble behaviors from the narrow side and wide side of the narrow rectangular channel. Bubble growth rate, bubble departure diameter, bubble interface parameter and others were obtained according to the observation. The bubbles always slide along the heating wall after departing from the nucleation sites. During the process of the sliding bubble motion along the heating wall, the upstream and downstream contact angle is almost equal. The phenomenon on bubble lift-off from the heating wall is not observed in vertical flow boiling with low heat flux in the isolated bubble region. The bubble tends to lift-off from the heating wall after sliding some distance in inclined upward facing upflow boiling and in vertical downflow boiling.

An analysis of force balance on a growing bubble is performed to analyze the mechanism of bubble departure, slide and lift-off behavior in the narrow channel. The bubble growth force is much less than the other forces acting on a bubble, and the buoyancy force, surface tension force, and quasi-steady drag force are the main forces controlling the bubble departure. The buoyancy, quasi-steady drag force, and added-mass force in the flow direction are the main forces controlling the sliding bubble motion along the heating wall. As the time increases, since the sliding bubble velocity exceeds the local liquid velocity of the center of mass of the bubble, and therefore the shear lift force in the y direction will push the bubble against the wall.

Acknowledgements

This work is supported by the National Natural Science Foundation of China (No.11475161 and 51106142).

Nomenclature

F_x	composition of forces in the x direction
F_y	composition of forces in the y direction
F_{qs}	quasi-steady drag force
F_b	buoyancy
F_{sl}	shear lift force
F_s	surface tension force
F_{am}	added-mass force
F_{cp}	contact pressure force
F_h	hydrodynamic pressure force
V_b	bubble volume
q	heat flux
G	mass flux
u	local liquid velocity of the center of mass of the bubble
v	bubble velocity
t	time
T	temperature
d_w	bubble contact diameter
τ_w	wall shear stress
g	gravitational acceleration
r	bubble radius

$a(t)$	bubble growth rate
Re_b	bubble Reynolds number
U	bulk velocity,
C_f	friction coefficients
λ	friction factor
D_h	hydraulic equivalent diameter
R	ratio of the bubble contact diameter to bubble departure diameter
C	empirical constant
Greek symbols	
a	upstream contact angle
β	downstream contact angle
θ	inclination angle
ρ_l	density of liquid,
ρ_v	density of bubble
γ	liquid kinematic viscosity
Φ	incline angle of the heating wall
Subscripts	
x	direction parallel to the heating wall
y	direction normal to the heating wall
b_x	direction parallel to bulk flow
b_y	direction normal to the heating wall
b_z	direction perpendicular to flow
in	inlet

Author details

Xu Jianjun*, Xie Tianzhou, Chen Bingde and Bao Wei

*Address all correspondence to: xujjun2000@sohu.com

CNNC Key Laboratory on Nuclear Reactor Thermal Hydraulics Technology, Nuclear Power Institute of China, Chengdu, China

References

[1] Peng XF, Wang BX. Forced-convection and flow boiling heat transfer for liquid flow through micro-channels. International Journal of Heat and Mass Transfer. 1993;**36**(14): 3421-3427. DOI: 10.1016/0017-9310(93)90160-8

[2] Mishimal K, Mibiki T, Nishihara H. Some characterstics of gas-liguid flow in narrow rectangular duct. International Journal of Multiphase Flow. 1993;**19**(1):l15-124. DOI: 10.1016/0301-9322(93)90027-R

[3] Tran TN, Wambsganss MW, France DM. Small circular- and rectangular-channel boiling with two refrigerants. International Journal of Multiphase Flow. 1996;**22**(3):485-498. DOI: 10.1016/0301-9322(96)00002-X

[4] Jiang L, Wong M, Zohar Y. Forced convection boiling in a micro-channel heat sink. Journal of Microelectromechanical Systems. 2001;**10**:80-87. DOI: 10.1109/84.911095

[5] Kandikar SG. Fundamental issues related to flow boiling in minichannels and microchannels. Experimental Thermal and Fluid Science. 2002;**26**(2–4):389-407. DOI: 10.1016/S0894-1777(02) 00150-4

[6] Piasecka M, Hozejowska S, Poniewski ME. Experimental evaluation of flow boiling incipience of subcooled fluid in a narrow channel. International Journal of Heat and Fluid Flow. 2004;**25**(2):159-172. DOI: 10.1016/j.ijheatfluidflow.2003.11.017

[7] Lie YM, Lin TF. Saturated flow boiling heat transfer and associated bubble characteristics of R-134a in a narrow annular duct. International Journal of Heat and Mass Transfer. 2005;**48**(25–26):5602-5615. DOI: 10.1016/j.ijheatmasstransfer.2005.05.013

[8] Lie YM, Lin TF. Subcooled flow boiling heat transfer and associated bubble characteristics of R-134a in a narrow annular duct. International Journal of Heat and Mass Transfer. 2006;**49**(13–14):2077-2089. DOI: 10.1016/j.ijheatmasstransfer.2005.11.032

[9] Cheng P, Wang GD, Quan XJ. Recent work on boiling and condensation in microchannels. Journal of Heat Transfer. 2009;**131**(4):1-15. DOI: 10.1115/1.3072906

[10] Yu W, France DM, Wambsganss MW, Hull JR. Two-phase pressure drop, boiling heat transfer, and critical heat flux to water in a small-diameter horizontal tube. International Journal of Multiphase Flow. 2002;**28**(6):927-941. DOI: 10.1016/S0301-9322(02)00019-8

[11] Chen DQ, Pan LM, Yuan DW, Wang XJ. Dual model of bubble growth in vertical rectangular narrow channel. International Communications in Heat and Mass Transfer. 2010; **37**(8):1004-1007. DOI: 10.1016/j.icheatmasstransfer.2010.06.023

[12] Chen DQ, Pan LM, Ren S. Prediction of bubble detachment diameter in flow boiling based on force analysis. Nuclear Engineering and Design. 2012;**243**:263-271. DOI: 10.1016/j.nucengdes.2011.11.022

[13] Xu JJ, He JS, Chen BD, Huang YP, Yan X, Yuan DW. Experimental visualization of sliding bubble dynamics in a vertical narrow rectangular channel. Nuclear Engineering and Design. 2013;**261**(8):156-164. DOI: 10.1016/j.nucengdes.2013.02.055

[14] Xu JJ, Chen BD, Xie TZ. Experimental and theoretical analysis of bubble departure behavior in narrow rectangular channel. Progress in Nuclear Energy. 2014;**77**(4):1-10. DOI: 10.1016/j.pnucene.2014.06.002

[15] Xu JJ, He JS, Chen BD, Wang XJ. Visualization of behavior of subcooled boiling bubbles in narrow rectangular slits. Journal of Power Engineering. 2007;**27**(3):389-392. DOI: 10.3321/j.issn:1000-6761.2007.03.020

[16] Cornwell K. The influence of bubbly flow on boiling from a tube in a bundle. International Journal of Heat and Mass Transfer. 1990;**33**(23):2579-2584. DOI: 10.1016/0017-9310(90)90193-X

[17] Thorncroft GE, Klausner JF, Mei R. An experimental investigation of bubble growth and detachment in vertical upflow and downflow boiling. International Journal of Heat and Mass Transfer. 1998;**41**:3857-3871. DOI: 10.1016/S0017-9310(98)00092-1

[18] Thorncroft GE, Klausner JF. The influence of vapor bubble sliding on forced convection boiling heat transfer. Journal of Heat Transfer. 1999;**121**(1):73-79. DOI: 10.1115/1.2825969

[19] Kenning DBR, Bustnes OE, Yan Y. Heat transfer to a sliding vapour bubble. Multiphase Science Technology. 2000;**14**(1):75-94. DOI: 10.1615/MultScienTechn.v14.i1.20

[20] Li X, Hollingsworth K, Witte LC. Vapor bubble rise under a heated inclined plate. Experimental Thermal and Fluid Science. 2007;**32**(2):529-544. DOI: 10.1016/j.expthermflusci.2007.06.003

[21] Levy S. Forced convection subcooled boiling–prediction of vapor volumetric fraction. International Journal of Heat and Mass Transfer. 1967;**10**(7):951-965. DOI: 10.1016/0017-9310(67)90071-3

[22] Klausner JF, Mei R, Bernhard DM, Zeng LZ. Vapor bubble departure in forced convection boiling. International Journal of Heat and Mass Transfer. 1993;**36**(6):651-662. DOI: 10.1016/0017-9310(93)80041-R

[23] Thorncroft GE, Klausner JF, Mei R. Bubble force and detachment models. Multiphase Science and Technology. 2001;**13**(3&4):35-76. DOI: 10.1615/MultScienTechn.v13.i3-4.20

[24] Situ R, Hibiki T, Ishii M, Mori M. Bubble lift-off size in forced convective subcooled boiling flow. International Journal of Heat and Mass Transfer. 2005;**48**(24–25):5536-5548. DOI: 10.1016/j.ijheatmasstransfer.2005.06.031

[25] Cho Yun J, Yum SB, Lee JH, Park GC. Development of bubble departure and lift-off diameter models in low heat flux and low flow velocity conditions. International Journal of Heat and Mass Transfer. 2011;**54**(15–16):3234-3244. DOI: 10.1016/j.ijheatmasstransfer.2011.04.007

[26] Zuber N. The dynamics of vapor bubbles in nonuniform temperature fields. International Journal of Heat and Mass Transfer. 1961;**2**(1):83-98. DOI: 10.1016/0017-9310(61)90016-3

[27] Zeng LZ, Klausner JF, Mei R. A unified model for the prediction of bubble detachment diameters in boiling systems—II. Flow boiling. International Journal of Heat and Mass Transfer. 1993;**36**(9):2271-2279. DOI: 10.1016/S0017-9310(05)80112-7

[28] Delnoij E, Kuipers JAM, van Swaaij WPM. Dynamic simulation of gas–liquid two-phase flow: Effect of column aspect ratio on the flow structure. Chemical Engineering Science. 1997;**52**(21–22):3759-3772. DOI: 10.1016/S0009-2509(97)00222-4

[29] Saffman PG. The lift on a small sphere in a slow shear flow. Journal of Fluid Mechanics. 1965;**22**(2):385-400. DOI: 10.1017/S0022112065000824

[30] Mei R, Klausner JF. Shear lift force on spherical bubbles. International Journal of Heat and Fluid Flow. 1994;**15**(1):62-65. DOI: 10.1016/0142-727X(94)90031-0

[31] Xu JJ, Chen BD, Wang XJ. Prediction of sliding bubble velocity and mechanism of sliding bubble motion along the surface. Journal of Enhanced Heat Transfer. 2010;**17**(2):111-124. DOI: 10.1615/JEnhHeatTransf.v17.i2.10

[32] Maity S. Effect of velocity and gravity on bubble dynamics [thesis]. University of California. 2000

Section 4

Radionuclide Migration in Environmental Systems

Chapter 6

Understanding Sorption Behavior and Properties of Radionuclides in the Environment

Roger Saint-Fort

Additional information is available at the end of the chapter

http://dx.doi.org/10.5772/intechopen.76215

Abstract

Prediction of fate and behavior of radionuclides in the environment is largely governed by sorption processes. Radionuclides physico-chemical species interacting with prevailing abiotic properties of the environment vary widely among varying constituting environmental components. Herein, this work discussed the most significant aspects of sorption processes and properties at the solid-water interface. Main sorption mechanisms were investigated using kinetic, thermodynamic analyses, and various mathematical models in current use for description of sorption–desorption processes in the environment. Knowledge of environmental transport, environmental pathways, and exposure pathways to radionuclides is also an important aspect of any strategy to protect the public and the natural ecosystems. In the final analysis, the choice of a functional sorption equation model will be dictated by the risk based under consideration, the level of information available, and the intrinsic accuracy of the predictive model.

Keywords: sorption, radionuclides, isotherms, kinetic, models, exposure pathways

1. Introduction

Different types of radionuclides or radioisotopes can be found in nature. They tend to be ubiquitous in the biosphere as they can occur naturally or deliberately synthesized. Radioactivity occurs as the result of a parent radionuclide spontaneously disintegrating and in the process forming a daughter nuclide by releasing gamma, beta and/or alpha radiations. Natural radionuclides are primordial, secondary or cosmogenic of origins. Artificial radionuclides are produced by nuclear explosions, nuclear reactors, particle accelerators or radionuclide generators. Radionuclides that find their way into the environmental spheres may cause harmful effects as radioactive contamination. Furthermore, they can also represent an occupational health risk if not properly managed.

Radiation may be released from electron excitation and emission. This process occurs by the electron absorbing energy from external sources of electromagnetic radiation. The extra energy absorbs by the excited electron is in due course released by emitting electromagnetic radiation of lower energy. Hence, the energy of an electromagnetic wave is directly proportional to its frequency and inversely proportional to the wavelength. Radionuclides intrinsically will have excess of energy due to nuclear instability. Radioactive decay arises because a nuclei is unstable. The ratio of neutron to proton and the total number of nucleons in the nucleus are the two determining factors that help ascertain the degree of stability of an isotope. However, to become more stable radionuclides emit particles or rays. The latter may be emitted as gamma radiation from the nucleus. Another possibility, the excess energy may be successfully transferred to its electrons or used to create and emit alpha or beta particles. Radiation is the release of energy particles and rays from atoms. Accidental release of radionuclides along with their decay products end up naturally in soil, water and air ecosystems. Many of these radionuclides can be anthropogenic generated and accidentally released in the environment leading to associated health risks [1]. United States Nuclear Regulatory Commission (U.S.NRC) reported that on an annual basis a person will typically receive an annual dose of 620 millirem. In that regard, natural sources account for 50% of radiation exposure that people generally receive while anthropogenic sources account for the remaining 50% (**Table 1**) [1].

Radionuclides such as radium-226, radium-228, thorium-232, uranium-238, and iridium-192 are commonly found in various ecosystems. **Table 2** depicts the details of the natural decay reactions for selected radionuclides [2]. The symbols α, β and γ indicate alpha, beta and gamma decay and the times shown are half-life. In nature, the radionuclides in some series are approximately in a state of secular equilibrium. This entails that that the activities of all radionuclides within a series are nearly equal. All radium in nature is radioactive. It is found in each of the three natural radioactive series-the thorium series, the uranium series, and the actinium series. However, radium higher solubility and its decay to radon makes it a primary environmental concern. As a radioactive gas, radon is colorless, odorless and tasteless. Inhalation is the primary route of radon exposure as it seeps into homes viz.; cracks in walls, floors, foundations and through floor drains and sumps.

Weathering releases and concurrent speciation of radionuclides and nuclides to water, sediment, soil and air with potential for further distribution and dispersion via dust and biota uptake has to be taken into account in modeling and assessing their mobilization and

Natural background	**Anthropogenic**
Space—5%	Consumer products—2%
Ecosystems—3%	Nuclear medicine—12%
Internal—5%	Occupational and Industrial—0.1%
Radon and Thoron—37%	Medical procedures—36%

Table 1. Typical sources and radiation exposure level.

Uranium 238 α ↓ 4.5 billion years	Thorium 232 α ↓ 14 billion years
Thorium 234 β ↓ Υ 24 days	Radium 228 β ↓ Υ 5.8 years
Protactinium 234 β ↓ Υ 24 1.2 min	Actinium 228 β ↓ Υ 6.1 hours
Uranium 234 α ↓ Υ 240,000 years	Thorium 228 α ↓ 1.9 years
Thorium 230 α ↓ Υ 77,000 years	Radium 224 α ↓ 3.7 days
Radium 226 α ↓ Υ 1600 years	Radon 220 α ↓ 56 seconds
Radon 222 α ↓ 3.8 days	Polonium 216 α ↓ 0.15 second
Polonium 218 α ↓ 3.1 minutes	Lead 212 β ↓ Υ 11 hours
Lead 214 β ↓ Υ 27 minutes	Bismuth 212 — — — α 61 minutes (64%) — > Polonium 212 α ↓ 61 minutes (36%) α ↓ 310 nanoseconds
Bismuth 214 β ↓ Υ 20 minutes	Thallium 208 ↓ β ↓ Υ 3.1 minutes ↓
Polonium 214 α ↓ 160 μ seconds	Lead 208 Stable
Lead 210 β ↓ Υ 22 years	
Bismuth 210 β ↓ Υ 5 days	
Polonium 210 α ↓ 140 days	
Lead 206 Stable	

Adapted from Argonne National Laboratory, EVS, human health fact, august 2005 [2]

Table 2. Natural decay series for the radionuclides of uranium 238 and thorium 232.

remobilization. More precise knowledge of the impact of various abiotic components are paramount. Interaction of radionuclides with various physical and/or chemical constituents present in environmental systems represents a key factor in affecting their environmental speciation and mobility. It has been reported that the migration and mobility of radioisotopes in the presence of mineral surfaces plays a vital role in predicting the environmental impacts in the case of an accidental release. Hence as vitally important in assessing the risk associated with radioactive waste disposal or radioactive contaminated sites [3].

The term adsorption refers to the adhesion of contaminants, termed adsorbate, to a surface, termed adsorbent. Hereafter, the term sorption is used to refer to the combined process of ad- and ab-sorption processes occurring at the surface of a sorbing material. Many investigations have suggested that sorption–desorption process is one of the most important factors affecting the fate and behavior of radionuclides and contaminants in the environment [4–7]. The process can occur at the interface between any two phases such liquid–solid, gas–solid, gas–liquid or liquid–liquid [8]. Varying factors have been reported to influence the process including organic matter content and degree of humification, dissolved organic matter, surface tension of aqueous phase, level of system pH, buffering capacity and ionic strength, type of clay and content, change in system redox, competition for sorbing sites, characteristics of sorbent and sorbate, precipitation and temperature fluxes, interaction mechanisms and contribution of different types of binding phase, sorbent solubility in a given system boundary, weathering as well as the concentration of the contaminant. Understanding how these interactive factors will allow better discerning and modeling of environmental pathways of radionuclides is germane. Therefore, effective strategy can be formulated to protect the health of at risk biota from harmful radionuclide contaminants.

In response to the potential environmental and health risks posed by the natural or anthropogenic release of radionuclides, this work explores and examines functional sorption models that in current use for modeling the behavior of radionuclides in soil-water phase systems.

2. Fundamental of radioactivity

All matter is composed of atoms. The atoms are known as the smallest indivisible particle that retains the characteristics of the element and that can take part in a chemical reaction. There are a reported number of 112 different elemental types of atoms. Atoms combine to form molecules. The molecule is the smallest particle of a substance that can consist of two or more atoms. The structure of a typical atom is shown in **Figure 1**. The atom may be described in terms of the numbers of three sub-atomic properties protons, neutrons and electrons. The protons and neutrons in the nucleus are collectively referred to as nucleons. The characteristics of each particle are summarized in **Table 3**. The nucleus is at the center of the atom containing protons and neutrons of approximately equal numbers. Because of this, all atoms tend to toward the stable of electrical neutrality. The nucleus of an atom represents almost the entire

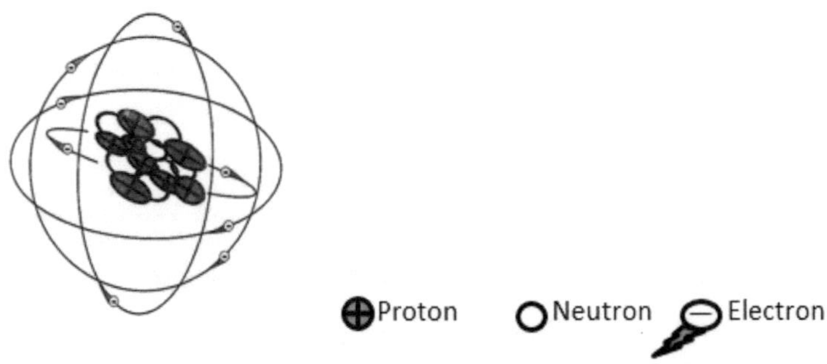

Figure 1. Structure of a typical atom.

Sub-atomic particles	Fundamental characteristics
Proton	• Applicable forces are coulombic, strong and weak nuclear • +1 charge • 1 amu* • Located in the nucleus • Symbol is p^+ or p
Neutron	• Applicable forces are strong and weak nuclear • 0 charge • 1 amu • Located in the nucleus • Symbol is n
Electron	• Applicable forces are coulombic • −1 charge • 5.5×10^{-4} amu • Orbits around the nucleus shell • Symbol e^- or e

*amu = atomic mass unit.

Table 3. Some key characteristics of sub-atomic particles.

mass of the atom. Compared to electrons, protons and neutrons are very heavy and will impart to the atom its mass.

As Rutherford and Bohr discovered, electrons can exist outside of the nucleus in discrete orbits or shells. Each will have a specific energy level. One of the ultimate consequences of quantum theory is that only a certain specific number of electrons can occupy each shell. Hence, the number of electrons in the outmost shell of an atom will define the way it can react with other atoms.

The nucleus is held by nuclear force that balances the attraction gravitational and the repulsive electric force. The interaction between strong attractive force and strong repulsive electrostatic force occurring between protons gives rise to the observed stability of the nucleus [9]. When dissociating into other elements, a consequence of that, radioactive elements are unstable, they release a large amount of energy. They undergo this decay process to achieve a state of smaller stable nuclei. Stability, the lowest energy state is achieved through the emission of different types of particles.

Atoms are arranged in the Periodic Table in order of increasing numbers of protons. The term nuclide is used to describe particular nuclear species with a given combination of A and Z characterized by the symbol ${}^{A}_{Z}X$. While X represent the symbol of the element, A and Z are the atomic mass number and the atomic number, respectively. Radioactive nuclides are referred to radionuclides. Atoms that share the same physical and chemical properties are referred to as elements. Isotopes are atoms of the same element that have a different number of neutrons but the same number of protons.

Many nuclei are radioactive; that is, they decompose by emitting particles and in doing so, become a different nucleus. An unstable atom will release energy in the form of radiation. In

Alpha Decay, α: ${}^{A}_{Z}X \;-->\; {}^{A-4}_{Z-2}X' + {}^{4}_{2}He^{*}$ $^{*}\alpha$ particles emission are mono-energetic	Due to their high charge, mass and energy, alpha radiation can potentially ionize tissue and cause biological damage.
Beta Decay, β: ${}^{A}_{Z}X \;-->{}^{A}_{Z+1}\; X'$+e- $+\bar{\upsilon}$ ${}^{A}_{Z}X \;-->{}^{A}_{Z-1}\; X'$+e+ $+\upsilon$ ${}^{A}_{Z}X$ + e- $-->{}^{A}_{Z-1}X' + \upsilon$❈ ❈Electrons capture process	Can penetrate human skin to alter chemical bonds and destroy living cells. Can also cause nausea, vomiting, dizziness, hair loss, cancer.
Gamma Decay, γ: ${}^{A}_{Z}X\bullet \;-->{}^{A}_{Z}X + \gamma$ ●Nucleus is in an excited state	Ionizing radiation can cause acute health effects and which can accumulate over time. Health effects can be bone marrow failure, cancer, immediate or delayed death.

Table 4. Comparison of the three types of radioactive decay.

natural radioactive decay, radiation is classified according to the nature of the particles emitted, the amount of energy associated with their emission, and the mechanism by which the emission arises. Three types of radioactive decay (**Table 4**) have been identified known as alpha decay (α), beta decay (β), and gamma decay (γ). Key factors have been highlighted in **Table 4.**

3. Radioactive decay equations

All radioactive material undergoes radioactive decay at a distinct rate essentially unaffected by variation in pressure, magnetic field, temperature or other external factors [9]. The rate of disintegration for any radionuclide is proportional to the original atoms present expressed as Eq. (1):

$$dN/dt = -ɋN \tag{1}$$

where dN/dt is the rate of change of quantity N relative to time, ɋ is the decay constant and the minus indicates a decrease in the quantity of N.

The solution to the above equation is given as Eq. (2):

$$N_t = N_o e^{-ɋt} \tag{2}$$

where N_t is the number of nuclides remaining at time t, N_o refers to the initial quantity of nuclides. According to Eq. (2), the amount of nuclides remaining at time t is occurring exponentially.

The time required for an amount of radioisotope to decrease by one-half is termed the radioactive half-life expressed as Eq. (3):

$$T_{1/2} = -\mathrm{Ln}(0.5)/K \quad (3)$$

or as Eq. (4):

$$K = 0.693/T_{1/2} \quad (4)$$

where K is as previously described.

Radionuclide decay is said to be isotropic meaning it is not directional. Because of such behavior, the decay of radionuclide can be mathematically represented by the inverse-square law. This entails that the flux of radiation emitted by a radioactive material is inversely proportional to the distance squared described by Eq. (5):

$$G = N/4\text{Л}d^2 \quad (5)$$

where G is the number of particles emitted per cross unit area, N represents the total number of particles emitted and D is the distance from the emitting source. An effective way to reduce radiation dose exposure is to maintain a safe distance.

4. Chemical characteristics of radionuclides

The presence of radionuclides in a given environmental system is fundamentally a transitory phenomenon. Environmental systems in the wider biogeochemical cycling are interconnected, dynamic and heterogeneously unique at the biotic and abiotic elements. Radionuclides are unstable isotopes of elements of metals and non-metals which undergo radioactive unlike heavy metals. It is well established that the transport behavior of radioisotopes or radionuclides and heavy metals, their bioavailability, their uptake in different food chains, and their toxicity are governed by chemical and physical properties, mainly speciation [10–12]. The term "speciation" often leads to technical ambiguity. In this work, speciation is defined in a broader sense as being the physico-chemical forms of a radionuclide or contaminant within an environmental system.

Radionuclides and nuclides released in the biosphere may be grouped according to their chemical properties (**Table 5**). It has been reported that their chemical behavior will tend to mimic stable elements in the same chemical group particularly the heavy metal elements. The chemistry of both radionuclides and heavy metals is far from being uniform with respect to their behavior and speciation relationship. Despite the paramount importance of speciation for their behavior in the environment, our knowledge remains scarce. At its most fundamental, very often the more extensive knowledge about the speciation of non-radioactive trace elements and heavy metals can be appreciably extrapolated to radionuclides [13]. An outline of selected forms of radionuclides is summarized in **Table 6** [14]. It has been reported that if the radionuclide is a chemical analogue of an essential nutrient, absorption of the radionuclide by biota will increase if the nutrient is scarce. Empirical knowledge of the chemical behavior of stable elements in the same chemical group is of paramount to assessing and gaining insights of the potential for uptake into food chains and exposure pathways.

Representative elements	Radionuclide types and atomic mass
Noble gases	Rn (222), Xe (131), Ar (39.95), Kr (83.80)
Heavy metals	Pb (207.2), Cr (52), Co (58.93), Ru (101.1), Mn (54.94), Zr 91.22), Po (209), Mn (54.94), Zn (65.38)
Non-metals	I (126.9), C (12.01), H (1.008), P (30.97)
Rare earths	Ce (140.1), Pr (140.9), Pm (145), La (138.9), Y (88.91)
Light metals	Ba (137.3), Ra (226), Ca (40.08), Cs (132.9), Rb (85.47), K (39.10), Sr. (87.62)
Actinides	U (238), Th (232), Pu (244) N.B. All actinides are radioactive.

Table 5. Radionuclides periodic table grouping according to chemical properties.

Speciation in	Forms	Characterization
Atmosphere media	Ionic Gaseous molecules Aerosols	Nature Nature Nature, radionuclide bonding
Water media	Ionic Ionic pairs Molecules Colloidal Particulate	Nature, simple, complex Nature Nature, inorganic, organic Nature, radionuclide bonding Substrate type, radionuclide bonding
Sediments, soils and rocks media	Dissolved forms in interstitial water or soil Sorbed Co-precipitated with	As above Molecules, ions, sorbent characteristics, bonding Carbonates, oxides etc. React and interact with organic matter Diffusion in sorbent lattices, fissure

Table 6. Main forms of radionuclides adapted from von Gutten and Benes [14].

5. Environmental and exposure pathways

Exposure routes are the ways radionuclides can enter a person. It can be by inhalation, ingestion or dermal uptake. The exposure route can be singular or multiple in an exposure event. While widely accepted, inhalation is the most rapid route of uptake, followed by dermal contact and ingestion. However, the manifested health effects may vary radically among the exposure routes. Common radiation units are summarized in **Table 7**.

Radionuclides release are either natural or accidental and can be considered as uncontrolled releases. In many scenarios, uncontrolled releases are under no direct control and are exemplified in major pollution incidents. Upon finding their way in their environment, radionuclide pollutants move and respond to a number of interrelated natural and anthropogenic factors. The interrelationship of these factors make the understanding of radionuclide contaminants fate and behavior fundamentally complex. Radionuclides are transported by the action of wind and water. However, their uptake in the biosphere is largely dictated by their speciation in a corresponding environmental system.

Quantity	Traditional unit	S.I. unit
Activity	Curie (Ci)	Becquerel (Bq)
Exposure	Roentgen (R)	Coulomb/Kilogram (C/Kg)
Absorbed dose	Rad	Gray (Gy)
Equivalent dose	Rem	Sievert (Sv)

Table 7. Radiation units and their equivalents.

A chain of events may occur that may expose human and ecological receptors. In that regard, when occurred, this chain is referred to as an environmental pathway. In this context, the exposure or environmental routes can be multiple concurrent and/or sequential and are the ways by which radionuclide agents can enter a receptor. The critical factors that define an exposure event must be understood and defined. The latter can be short term or long term duration. Noteworthy is that the characterization pathway framework should consist of the following elements and preferably ranked by their impact on health (**Table 8**) [15]. For obvious economic reasons, the process should be first and foremost risk-based.

The introduction of radionuclides into an ecosystem can provide a direct or indirect hazard to individual receptors. An exposure route is the manner that a contaminant of concern enters a receptor [16]. Exposure could be via inhalation, considered as the most rapid route of uptake, followed by dermal contact and ingestion, respectively. Exposure period can be chronic, sub-chronic or acute [17]. The latter type of exposure causes death. On the other hand, chronic exposure occurs over a long period of time, and negative health effects are cumulative. There are many other links among environmental media, exposure media and exposure pathways scenarios. Evidently, they may be constructed depending on the specific needs of the risk-based assessment. A generalized exposure model is illustrated in **Figure 2**. It is also important to establish potential pathways between sources and receptors for each radionuclide of concern.

Elements	Remarks
Radionuclides properties and concentration	Radiation hazard
Release mechanisms	Leaching
Time and space scale of radionuclide concentration	Aerial distribution and natural attenuation
Exposure duration	Size and nature of the population
Dose–response	Dose/response adverse health effect
Transfer mechanisms	Sorption–desorption
Transformation mechanisms	Stable isotopes, physical state change
Exposure point	Residential well
Exposure route	Inhalation
Receptors	Age, Gender
Contributing environmental media	Air, soil, lagoon

Table 8. Framework of an exposure event analysis.

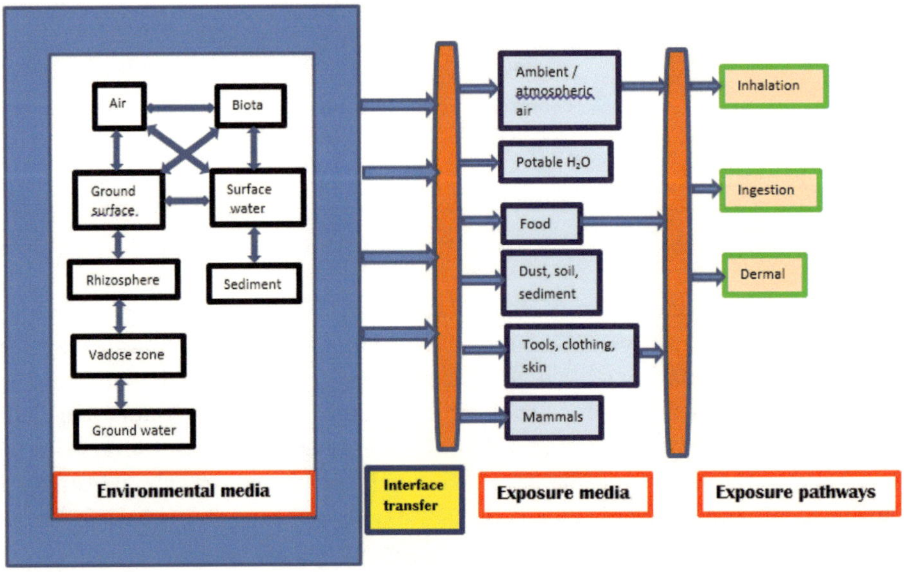

Figure 2. Generalized exposure model.

6. Sorption theoretical concepts

Links among environmental media, exposure media and exposure pathways as well the interactive effects that they exert are greatly influenced by the process of sorption in a given scenario. Evidence of this was illustrated in research reported by Konoplev [18], Pfingsten et al. [19], and Verkhovskaya et al. [20] on the behavior of radionuclides in the environment. However, the emphasis of the sorption concepts will be on radionuclides-liquid–solid (i.e., soil, sediment systems). In considering such systems, the migration of radionuclides ions and colloidal particles will be dictated by movement of fluid and geochemically, being represented by sorption (i.e., adsorption ↔ desorption ↔ diffusion) behavior.

It is well known that sorption behavior of radionuclides in the environment, their toxicity and uptake in the biosphere are primarily dictated by physico-chemical properties, namely their speciation and to a lesser degree by their concentration. Speciation in the context of this work refers to the physico-chemical forms of radionuclides. Their primary speciation is highly function of its origin. According to Gutten and Benes [14], the sorption and desorption behavior of nuclides are strongly affected by the physical, geochemical and biological processes that are occurring at specific field sites. They further concluded that the mobility of a nuclide in an aquifer system depended strongly on the respective negative, neutral or positive charges and the nature of mineral surfaces and dissolved species. Complexation or variations in pH and/or redox state may change the identity of the primary species. Despite of a large number of investigations, our knowledge of the sorption behavior of nuclides in the environment is still limited.

Sorption occurs either as a physical adsorption, a phenomenon referred to as physisorption or chemical adsorption, also called chemisorption. Sorption of radionuclides can occur at the interface between any two phases in the environment between liquid–solid or gas–solid. Scientists agree that sorption is a complicated process. It depends largely on the composition of the fluid phase, surface chemistry of the host sorbate, the system pH and Eh and chemical speciation of radionuclides. Both repulsive and attractive forces become balanced when sorption occurs. Much definitive work relating to sorption of radionuclides have been published [21–24]. Mechanisms of sorption in many cases have been established and some relationships between structural properties of sorbate and sorbent have been investigated [25–27]. Behavior of radionuclides can then be correlated to stable elements in the same chemical group with respect to their sorption. Consequently, sorption mechanisms can be readily deduced accordingly. The literature relating to mechanisms of radionuclides sorption indicates that the process is based on the attraction force between the individual molecules, atoms or ions of a sorbate and the sorbent surface [28, 29].

7. Sorption mechanisms

Sorption mechanisms proposed by various scientists as being responsible for retention of pollutants and by extension radionuclides, in the environment include the following: (1) ion exchange, (2) physical or van der Waals' forces, (3) chemisorption, (4) coordination complexes or specific and (5) multicomponent sorption. Two or more mechanisms may operate simultaneously [30]. It has been reported that electrostatic charge generated by radioactivity contributes significantly to surface interactions between particles and environmental surfaces [31]. In assessing the potential for interactions with complexing agents, uptake into the food chain and mobilization, paradoxically the same governing variables and processes affecting heavy metals and metalloids in the biosphere can be extrapolated to analogue radionuclides [32, 33]. This leads to the discussion of sorption mechanisms:

7.1. Ion exchange sorption

This type of sorption is driven by the attractive force for maintaining neutrality. The electric charges of the surface of the sorbent are balanced by equivalent ions on opposite charge. Charge of the radionuclides is the determining factor for exchange sorption. It has been shown that radioactive particles can be strongly charged and yield an asymmetric bimodal charge distribution [34, 35]. Therefore the processes involving electrostatic interactions will play a role in predicting their transport. Undoubtedly, the nature of the charge will be a determining factor in their distribution in environmental systems. Generally divalent cations are more strongly attracted toward a site of opposite than monovalent cations by a matrix constituent such as soils or sediments. In the process of ion exchange, ionic radionuclides compete for the exchange sites. As such, a previously sorbed ion of weaker affinity can be exchanged for a radionuclide free ion sorbent of opposite charge [36]. Cation exchange sorption can be illustrated as follows:

where M^+ represents a positive ion replaced by a cation on the exchange complex. A key point is that ion exchange of metallic ions with soil and other matrices may be partially reversible in response to pH changes and contaminant concentrations [37, 38]. Sorption via this mechanism constitutes a retardation rather than attenuation process. Attenuation processes that can transform nuclides are hydrolysis, volatilization which results in transferring a nuclide to the atmosphere or oxidation-reactions initiated via chemical or biological pathways.

Acid–base equilibria appears to play an important role in sorption behavior of uranium, thorium and radium by calcite and dolomite matrices investigated under toxic environments [39]. Thorium sorption in the calcite and dolomite systems was very significant irrespective of the observed pH changes. It appears that the hydrolyzed species were preferentially sorbed on the surfaces.

7.2. Physical sorption

It is generally agreed that physical or van der Waals' forces operate in all molecules, but substantially weak. This type of electrostatic interaction between atoms and molecules arises from the fluctuations in their electron density and expected to also operate in nuclides. Such fluctuations produce dipole moments which contribute to sorption. They operate in all sorbate-sorbent relationships. It is also highlighted that sorption created by physical forces are additive meaning that each atom of a molecule of a nuclide or a sorbent contributes to the total bond energy. The attraction of two molecules resulting in a dipole–dipole interaction is illustrated as follows:

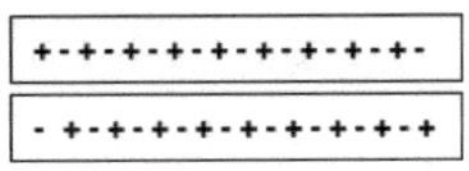

The individuality of the sorbate and the sorbent is always preserved. Physical sorption is always an instantaneous and reversible process requiring no activation energy and does not depend on the chemical nature of the sorbent.

7.3. Chemisorption

This involves that the nuclide undergoes a direct formation of chemical interaction with the sorbent. The phenomenon is sometimes referred to as activated sorption or chemical sorption. Chemisorption can occur at high temperature where physical sorptive processes would be less favored. A distinguishing feature of this type of sorption is the fact that sorption of contaminants can occur at extremely low sorbate concentrations and still able to saturate sorbent sites. When chemisorption occurs, the sorbed species are not able to move about the sorbent surface. Sorbates will usually occupy specific sorbing sites on the surface and form a monolayer of chemisorbed of contaminants [40]. Because chemisorption processes exhibit high energies of activation, chemical nature of the sorbate may be permanently altered.

7.4. Coordination complexes or specific

Fundamentally every aspect of the chemistry of heavy metals in the environment involves the formation of complexes with organic matter namely, humic and fulvic acids. The process in which a heavy metal cation combines with molecules or anions containing free pairs of electrons is known as coordination complex or specific sorption. Similar interaction can be postulated for nuclides. Knowledge of the distribution of radionuclide speciation in the environment is important for consideration of their behavior in various environmental systems. The anions or molecules with which a cation or nuclide forms a complex is generally referred to as a ligand. The stability of the ligand must be ascertained in order to understand its behavior in the environment [41, 42]. Stability constants can be used to elucidate this consideration. At equilibrium, this relationship can be represented as Eq. (6):

$$HM^{++} + 2L^{-} \leftrightarrow HML_2 \tag{6}$$

where HML_2 describes a coordination complex consisted of the association of the cation HM^{++} with the ligand L^{-}. The relative stability of the coordination complex HML_2 is expressed as Eq. (7):

$$K = [HML_2]/[HM^{++}][L^{-}]^2 \tag{7}$$

where the brackets indicate the concentrations of HML_2, HM^{++} and L^{-}, respectively, in moles L^{-1}. The significance of K constant is the larger the value the greater the stability of the ligand.

It has been found that coordination complexes exhibit a wide range of binding energies. This type of sorption often involves organic molecules whereby specific interactions between structural constituents of the sorbate and sorbent arises [42].

7.5. Multicomponent sorption

Competitive sorption between sorbates have been recognized as an important factor in determining contaminants migration in the environment. Multicomponent sorption phenomenon has also been reported in industrial applications [43–45]. It implies the sorption of the sorbed material to be will be a mixture of many compounds rather than a single one. Since the contaminants are competing for the open sorbing sites, several scenarios may be envisioned. The interactions may lead to mutual synergistic sorption, interfere with the sorption of each other, behave relatively independent. In multiple component systems of sorption, the sum of the desorbed and input concentration for each contaminant becomes an adimensional concentration. Adriana et al. [46] reported for multiple component systems, a large quantity of interaction sites are available on the sorbent and all the contaminants in the solution mixture easily sorbed. However, the contaminants with the lowest affinity for the active sites are desorbed from the sorbent and replaced by sorbates with a higher affinity. Similar empirical findings have been reported elsewhere [47–49].

8. Sorption kinetic models

Long- or short-term dynamics of radionuclides can be approached from a standpoint of basic mechanisms of nuclides sorption–desorption processes. Therefore, the ease in which nuclides move through the environment and are taken up by biotic and abiotic environmental compartment is governed by their chemical speciation and forms as well by site-specific environmental characteristics. However, limiting the impact of introduced radionuclides is in reality a very complex and difficult to resolve practically as illustrated by the various interconnectivity between various environmental components (**Figure 3**). In addition, toxicity of radionuclides appears to arise from three sources in the environment meaning the atmosphere, the hydrosphere and the lithosphere. However, hydrosphere would represent the largest source because it occupies the 75% of the earth's surface. In this context, corresponding models were formulated to describe the kinetics of sorption of contaminants in the various environmental compartments for liquid–solid phase systems. Reaction kinetics have been applied in the interpretation of the interactions of contaminants including radionuclides in the environment. In a general sense, the mobilization of a reactive nuclide will be dependent on the rate of the sorptive-desorptive reaction between the matrix solution and the solid phase. Travis and Etnier [50] reported such reaction process can either a kinetic one leading to a constant change in the relative amount of reactive sorbate in a solution matrix and in the solid phase ever changing as a function of time. Or, the equilibrium situation in the above relationship is quickly attained and therefore remain constant. In this situation, a concise definition of an equilibrium sorption–desorption reaction for a reactive nuclide entails that the rate of sorption between the matrix solution and the solid matrix is significantly greater than the rate of observed change in solute concentration in the matrix solution. According to Travis and Etnier [50] a kinetic situation is one for which an equilibrium sorption isotherm is not applicable.

The problem of understanding the various sorption mathematical relationships is of paramount for the purpose of evaluating the environmental behavior, managing remedial activities,

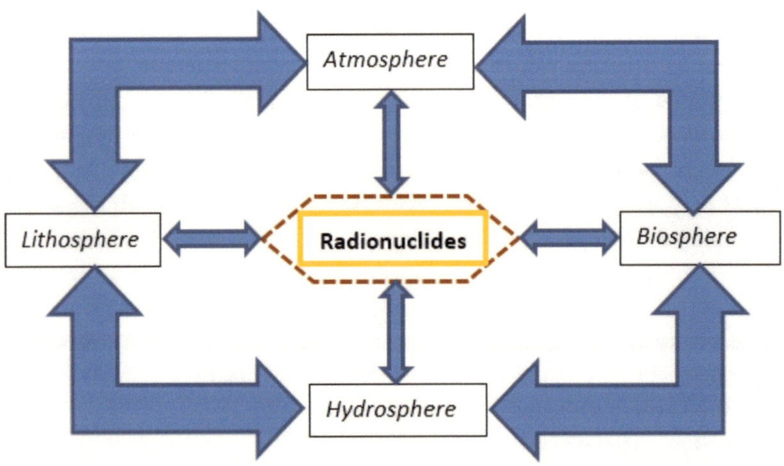

Figure 3. Interconnectivity of radionuclides in the environment.

bioavailability, accumulation and toxicity of nuclides and anthropogenic contaminants accidentally released. Nonetheless, an absolute predictive simulation capability may never be attained given the overall general complexity this entails.

Movement of a reactive radionuclide to a sorbent particle surface site would necessitate four distinct separate transport phenomena. These steps are initial physical attachment, then bulk fluid transport, film transport follows, and lastly intraparticle (or pore) diffusion. Film and intraparticle are normally viewed as the slowest and therefore the rate-limiting steps controlling sorption process. At equilibrium, there is a distinct distribution of sorbate between the fluid phase solid sorbent phases. Consequently, one performs an experiment where a specified mass of sorbent is equilibrated at a specific temperature and pressure with a known volume of specific concentration. Subsequently, the resulting equilibrium concentration in solution measured quantity sorbed is determined by difference. The empirical information is referred to as an adsorption isotherm. To allow for comparison of different sorbent materials, the quantity sorbed is always normalized by the mass of sorbent used in the experiment.

Several kinetic models have been reported to describe these sorption processes [51–54]. There are several approaches that are utilized to account for sorption–desorption processes on solid surfaces for liquid–solid interaction systems. These various kinetic models are subsequently discussed in the following sections.

8.1. Pseudo-first-order model

Lagergren [55] was first to present the pseudo-first-order model. It was used to describe the sorption of oxalic acid and malonic acid onto charcoal. The pseudo-first-order model works best for the initial 30 minutes of the sorption process. It is best expressed mathematically as Eq. 8:

$$\frac{dqt}{dt} = k(qe - qt) \tag{8}$$

where qe and qt are the sorption capacity at system equilibrium and at time t, respectively (mg sorbent/g sorbate) and k represents the rate constant sorption (time^{-1}). After integration and considering the boundary conditions $qe = 0$ at t = 0 and qt at t = t, Eq. (8) is rearranged and expressed as Eq. (9):

$$\text{Log}(qe - qt) = \text{Log}\, qe - \frac{kt}{2.303} \tag{9}$$

A linear regression analysis of Log $(qe - qt)$ vs. f(t) allows to derive the constant k as being the slope of the regression equation.

8.2. The Elovich model

This equation was developed by Roginsky and Zeldovich [56] but now generally known as the Elovich equation. It was based on a system sorption capacity. The Elovich equation is of the form Eq. (10):

$$\frac{dq}{dt} = A\exp(-bq) \tag{10}$$

where q represents the sorption capacity at time t (mg/g), A and b are parameters to the initial sorption rate (mg/gmin) and the desorption constant (g/mg).

An Elovich type of equation was derived by Chien and Clayton [57] by assuming a, b, t ≥ 1, thus arriving at Eq. (11):

$$qt = \frac{1}{b}\ln(Ab) + \frac{1}{b}\ln(t) \tag{11}$$

where A and b are obtained from the linear regression analysis of the function qt versus f(t).

8.3. Kinetic product model

This kinetic model was proposed Enfield [58]. The equation is of the form, Eq. (12):

$$\frac{dS}{dt} = aC^bS^d \tag{12}$$

where a, b, and d are parameter constants, S is the amount of contaminant sorbed, and C is the concentration of solute in in the matrix solution. This model does not imply a maximum sorbing capacity. It has been applied to describe phosphorous sorption and migration in the vadose zone [58].

8.4. Reversible linear model

The reversible kinetic linear model has been used to describe the sorption of solute [59, 60], Eq. (13):

$$\frac{dS}{dt} = \frac{\rho\Upsilon}{BD}(C - \beta S) \tag{13}$$

where S represents the amount of sorbate sorbed, C, the concentration of sorbent in solution, Υ, soil volumetric water content, BD, soil bulk density, while ρ and β are parameter constants. According to Eq. (13), the rate of sorbate sorption by the sorbent is proportional to the amount of sorbing sites available during the sorping interaction process. In some instances, transport of radionuclides may not be significantly enhanced by this process [61].

8.5. Pseudo-second-order

The equation for the pseudo-second-order is based on sorption equilibrium capacity [62–64]. The equation is written as Eq. (14):

$$\frac{dq}{dt} = a(q_1 - q_t)^2 \tag{14}$$

where a represents the second order rate constant. The constant a can be obtained by both linear and nonlinear regressions. The linear regression analysis is obtained by plotting t/q_t as a function of t.

8.6. Film diffusion model

A film diffusion model was derived by Mohan et al. [65]. It is governed by Eq. (15):

$$\mathrm{Ln}\frac{c}{co} = -ks\frac{S_s}{\varphi}t \tag{15}$$

where *ks* is the coefficient of diffusion, S_s is the external surface area of the sorbent while t presents time, and φ is the solution volume. The steps involved in film diffusion sorption processes are depicted in **Figure 4** [66].

According to Mohan et al. [65], several key conditions will govern the rate of sorption by film diffusion: (1) sorbent particle size (2) effective mixing (3) degree of sorbate affinity for sorbent surface and (4) relatively low sorbate concentration.

8.7. Mass transfer model

Also referred to as the liquid phase driving force (LPDF) [67]. In the mass transfer model [68], the rate of sorbate sorption is given by Eq. (16):

$$\frac{dS}{dt} = \mathrm{k}(\mathrm{C_o} - \mathrm{C_1}) \tag{16}$$

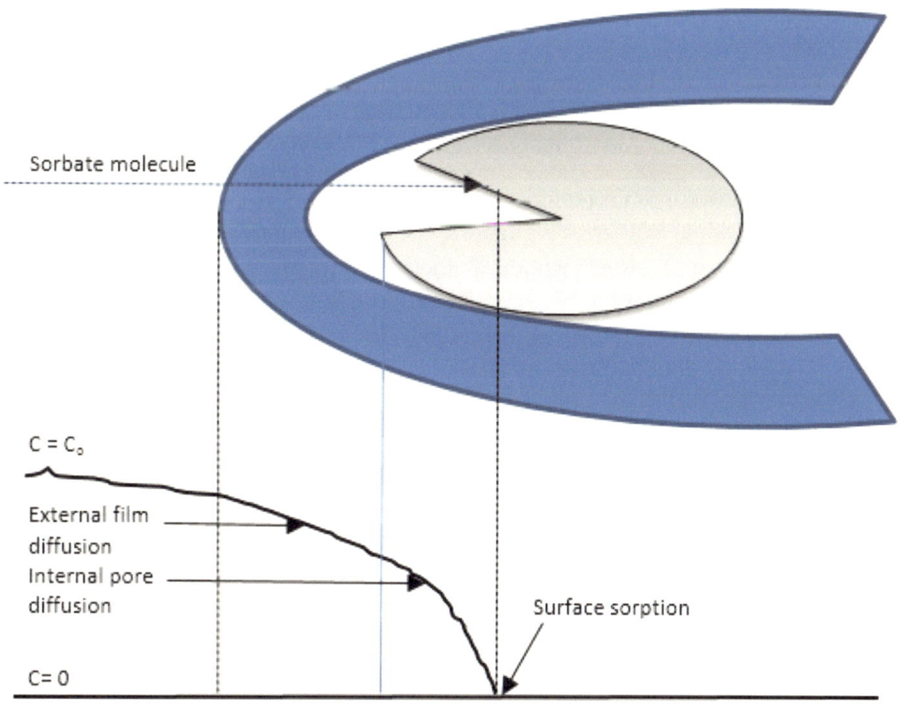

Figure 4. Film diffusion model adapted from Mohan et al. 2001 [65].

where k represents a sorption rate parameter describing the diffusive transport of the sorbate through a liquid layer surrounding a sorbent particle, C_o is the concentration of the sorbate in the solution phase, and C_1 is the solution phase concentration of the sorbate in instantaneous contact with the sorbent surface. The kinetic mass transfer model has been applied by several scientists [69–72].

8.8. Weber-Morris intra-particle model

According to this kinetic version model [73], if intra-particle diffusion limits the process, uptake would be governed by Eq. (17):

$$\mathfrak{p}_t = k_i\sqrt{t} + C \tag{17}$$

where $\mathfrak{p}_t$ is the amount of sorbate sorbed by the sorbent, k_i represents intra-particle diffusion rate constant, t is the time, and C is a constant related to resistance offered by boundary layer. This kinetic model has been applied by several researchers [74–76].

9. Sorption equilibrium models

Although limited studies have shown that sorption of nuclides to mineral and organic surfaces is a two-step process: a very rapid adsorption of radium to the external surface of the sorbent is followed by slow intraparticle diffusion along the micropores walls. Inspection of a microscopic cross-section of soil, organic matter or sediment particle reveals a porous structure with internal surface area that can vary from small to large. For the most part, this intraparticle diffusion can be regarded as one of the important rate-limiting mechanisms in the sorption of nuclides.

Construction of functional isotherms expressing the quantity of nuclides sorbed per unit mass of sorbent is referred to as an adsorption isotherm. However, equilibrium conditions must prevail and secondary reactions such as precipitation must be eliminated or corrected for. It exists several different mathematical forms of isotherms. Compliance with either the Freundlich, Langmuir, Polanyi-Dubinin-Manes or Dubinin-Radushkevich adsorption equations is based on empirical fittings the data indicating that the process either does or does not fit. The Freundlich isotherm is the most commonly used model. These isotherms have been broadly classified according to initial slope into four general isotherm classes according to **Figure 5** [77]. The L-curve (for Langmuir) is the most common. It indicates a relatively high affinity between the sorbent and sorbate in the initial stages of the isotherm. The slope becomes gradually less steep as sorbing sites become occupied. The S-curve indicates cooperative adsorption indicating that the sorbate has a high affinity for the solvent. The C-curve (constant partition) represents constant partition between solution and surface. This suggests that new sites become available as the sorbent is adsorbed. The H-curve characteristically occurs when the sorbate has a very high affinity for the solid.

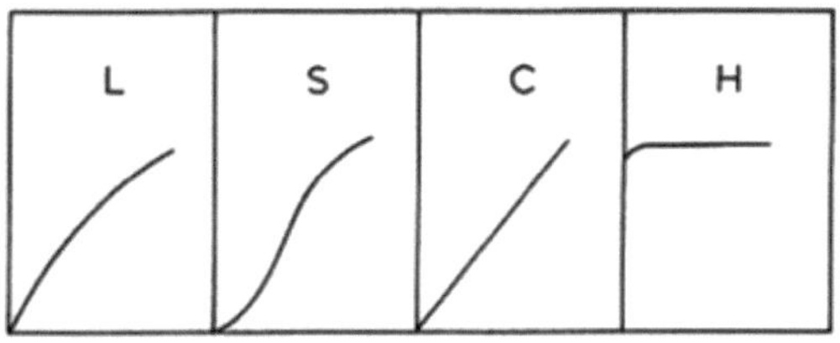

Figure 5. Classification of adsorption isotherms according to Giles et al. [78].

9.1. Langmuir isotherm model

The Langmuir equation was originally derived for the adsorption of gases by solids [78]. It is a semi-empirical isotherm and developed on the basis of assumptions that (1) the free energy of adsorption is a constant that is independent of the surface coverage (i.e., homogeneous surface), (2) there are no interactions between adsorbed molecules, (3) each site can accommodate one adsorbate molecule (i.e., monolayer adsorption), and (4) all sites are energetically equivalent. The common model form is Eq. (18):

$$\frac{X}{M} = \frac{KCeqb}{1 + KCeq} \tag{18}$$

where $\frac{X}{M}$ is the weight of sorbate per unit weight of sorbent, K is a constant related to the binding energy strength, Ceq is the equilibrium concentration of the sorbate, and b is the maximum amount of sorbate that can be adsorbed, meaning a complete monolayer surface coverage when $Ceq = \frac{1}{b}$ and $\frac{X}{M} = \frac{K}{2}$.

The Langmuir equation can be rearranged to the linear form Eq. (19):

$$\frac{Ceq}{X/M} = \frac{1}{Kb} + \frac{Ceq}{b} \tag{19}$$

If the empirical data conforms to the Langmuir equation, a plot of $\frac{Ceq}{X/M}$ versus Ceq yields a straight line with a slope $\frac{1}{b}$ and intercept $\frac{1}{Kb}$. The good fitting of the experimental does not necessarily all the underlying have been strictly met. The Langmuir equation model is limited to the range for which experimental has been derived. Extrapolation to higher concentration may yield inaccuracy in predicting sorption maximum for a system under investigation. Nonetheless, the value of b represents a practical limiting sorption capacity.

9.2. Freundlich isotherm model

The Freundlich isotherm model is a purely empirical model with an exponential distribution of site energies and immobile adsorption [79]. The Freundlich equation is often successfully used when adsorption data do not conform to the Langmuir equation. It infers that there exists multiple sets of heterogeneous sites on the surface of the sorbent for interaction to occur with the sorbate. The equation is expressed as follows, Eq. (20):

$$\frac{X}{M} = \mathrm{K_f}\, Ceq^{1/n} \tag{20}$$

where $\frac{X}{M}$ is the equilibrium sorbed concentration, Ceq is the equilibrium concentration of the sorbate, $\mathrm{K_f}$ is the Freundlich constant related to the adsorption capacity, and $1/n$ is the intensity of the adsorption. The equation implies that the energy of adsorption decreases logarithmically as the fraction of surface covered surface increases. Consequently, this limits Freundlich equation model to concentrations below saturation where sorption interaction would no longer be significant.

A linear form of the Freundlich equation is used in the analysis of experimental, Eq. (21):

$$\log \frac{X}{M} = \log \mathrm{K_f} + \frac{1}{n} \log Ceq \tag{21}$$

If $log \frac{X}{M}$ is plotted as a function of $logCeq$, a straight line should be obtained with an intercept on the ordinate of log $\mathrm{K_f}$ and slope $\frac{1}{n}$ on the graph. Intercept value of $\mathrm{K_f}$ is simply a measure of the distribution of the sorbate between the two phases and sometimes referred to as distribution coefficient, $\mathrm{K_d}$. As a reliable value, $\mathrm{K_d}$ is more accurate when the Freundlich $\frac{1}{n}$ approximate 1. The parameter $\mathrm{K_d}$ has sometimes been used in correlation studies for determining the relative importance of the various soil parameters on sorption. For a given sorbent, the parameter $\mathrm{K_d}$ is often normalized for the organic carbon of the system, $\mathrm{K_{oc}}$, as sorption in soils is highly correlated with soil organic matter content.

9.3. Polanyi-Dubinin-manes (PDM) isotherm model

The Polanyi-Dubinin-Manes (PDM) isotherm model [80–82] provides a theoretical framework for assessing sorption equilibrium capacities of a system. The model is written in the form of Eq. (22):

$$\log \mathrm{G_u} = \log G^s + \mathrm{a}(2\,\mathrm{V_e})^{\mathrm{n}} \tag{22}$$

where Eq. (23) defines 2 as

$$2 = \mathrm{RT} \ln \left(\frac{Ce}{Cs}\right) \tag{23}$$

and where the variables in the above equations are $\mathrm{G_u}$ is the amount sorbed at equilibrium, G^s is the system sorption capacity, a and n are constant parameters, 2 is the effective sorption potential, $\mathrm{V_e}$ is the molar volume of sorbate, R is the universal gas constant, T is the absolute temperature, $\mathrm{C_e}$ is the sorbate concentration at equilibrium, and $\mathrm{C_s}$ is the aqueous water solubility. In effect, the PDM assumes it exists a finite and fixed space associated with the sorbent for sorption to occur.

9.4. Dubinin-Radushkevich isotherm model

The model has the form of Eq. (24) [83]:

$$\text{T}_m = \text{CƆ}_b \exp\left(-\text{Ə}\text{Ʈ}^2\right) \tag{24}$$

where

$$\text{Ʈ} = \text{RT ln}\left(1 + \left(\frac{1}{Ce}\right)\right) \tag{25}$$

and where the variables in the above equations are T_m, the equilibrium sorption capacity, CƆ_b, the maximum sorption capacity, C_e, the equilibrium concentration, Ə, the binding energy constant, and Ʈ, the Polanyi potential.

The linear form is given by Eq. (26) allows to obtain the parameter Ə:

$$\text{E} = \frac{1}{\sqrt{2\text{Ə}}} \tag{26}$$

from which the free energy is then obtained.

In a general sense, the retention of radionuclides by solid solution cannot be differentiated from sorption in the environment. They are simply different explanations for the same phenomenon.

10. Effects of sorption on radionuclides migration

Sorption of radionuclides will have for implications to retard their transport in the vadose and saturated zones. Darcy's law is valid for unsaturated flow as water the transporting fluid. Hydraulic conductivity in the vadose zone is a function of moisture content of the soil. Darcy's law can be written as Eq. (27) [84]:

$$\text{V} = -\text{K}(\theta)\frac{\partial h}{\partial z} \tag{27}$$

where V is the Darcy velocity, K(θ) is the unsaturated hydraulic conductivity coefficient, θ is the volumetric water content, z is the soil depth below ground surface, h is the total head (z + ψ), ψ and is the soil tension or suction. The K value can be determined experimentally or in-situ [84].

The conceptual model for predicting transport of radionuclides in the vadose can be modeled with a simple source one-dimensional model. More sophisticated predictive models have been proposed elsewhere [85, 86]. In the application of one-dimensional model, the soil medium is assumed to be homogeneous, isotropic, and plug flow occurs. In the latter case, advection dominates and dispersion is considered negligible. Basically, infiltrating and percolating water at a constant velocity, $V\alpha$, is flowing through the vadose zone and carrying dissolved radionuclides at a specific concentration, $C\alpha$. This component can be sorbed by the soil or biodegraded by the microbial communities. The concentration of a radionuclide will be a function of both depth and time. The advection-dispersion equation can be written as follows [87], Eq. (28):

$$\frac{\partial C\alpha}{\partial t} = D_\alpha \frac{\partial^2 C\alpha}{\partial X^2} - V_\alpha \frac{\partial C\alpha}{\partial X} - \frac{p}{\theta}\frac{\partial Cs}{\partial t} - \mu_\alpha C_\alpha - \frac{p}{\theta}\mu_s C_s - KC \tag{28}$$

where $C\alpha$ is the concentration of the radionuclide, t is time, $D\alpha$ is the dispersion coefficient, X is the depth, p is the soil bulk density, θ is the volumetric water content of the soil, Cs is the concentration of the radionuclide in the solid phase, μ_α is the decay coefficient in the solution phase, μ_s is the decay coefficient in the soil phase, K is the constant decay, and $V\alpha$ is water flowing through the soil at a constant velocity. The equation model, Eq. (27) assumes that sorption is linear, hence Eq. (29) can be obtained:

$$C\alpha = K_d C \tag{29}$$

where K_d is the partition coefficient constant for the radionuclide derived from the equilibrium isotherm and C is radionuclide equilibrium concentration. In practice, the higher a K_d value the greater the relatively less mobile will be a radionuclide in the vadose or saturated zone. The partition coefficient K_d is more a function of the hydrophobicity and hydrophilicity of a contaminant. Accounting for the retardation factor that is the measure how much slower a radionuclide migrates than water, Eq. (28) may be rewritten in the form of Eq. (30) [88]:

$$\frac{\partial C\alpha}{\partial t} = \frac{D\alpha}{R}\frac{\partial^2 C\alpha}{\partial X^2} - \frac{V\alpha}{R}\frac{\partial C\alpha}{\partial X} - \frac{\mu C_\alpha}{R} - KC \tag{30}$$

where variables are as defined in Eq. (28) and R is the retardation factor. The retardation factor, R and current decay coefficient, μ can be derived and is given by Eqs. (31) and (32), respectively [88, 89]:

$$\mathrm{R} = 1 + \mathrm{p}\frac{Kd}{\theta} \tag{31}$$

where R is the retardation factor, p is the bulk density, K_d is the partition ratio, and θ is the volumetric water content of the soil matrix. Therefore,

$$\mu = \mu_\alpha + \mu_s \mathrm{p}\frac{Kd}{\theta} \tag{32}$$

where μ is the current decay coefficient, μ_α is the decay coefficient in the solution phase, μ_s is the decay coefficient in the soil phase, p is the bulk density, K_d is the partition ratio, and θ is the volumetric water content of the soil matrix.

Radionuclides mobilized in the vadose can be advectively transported and eventually reaching the groundwater. For a saturated zone, the one-dimensional of the advection-dispersion equation can be expressed as Eq. (33) [90, 91]:

$$\mathrm{R}\frac{\partial C}{\partial t} = D_L \frac{\partial^2 C}{\partial L^2} - V_\alpha \frac{\partial C}{\partial L} + \frac{BD}{\theta}\frac{\partial S}{\partial t} - \vartheta C \tag{33}$$

where C is the concentration of the radionuclide, t is time, θ is the porosity of the soil, V_α is the groundwater interstitial velocity, BD is the bulk density of the groundwater porous medium, L is length of plow path, S is the amount of sorbed radionuclide, and ϑ is the constant decay. In the groundwater, radionuclide movement would be primarily driven by advection, depicted in Eq. (34) [88]:

$$V_x = \frac{K}{n}\frac{dh}{dL} \tag{34}$$

where V_x the radionuclide interstitial velocity due to advection, K is the hydraulic conductivity, n is the porosity of the groundwater formation, and $\frac{dh}{dL}$ is the hydraulic gradient. The equation for predicting velocity and therefore travel time, can be written as follows using the partition coefficient described by Eq. (35):

$$V_x = K\frac{dh}{dL}\left(\frac{1}{\theta R}\right) \tag{35}$$

where V_x is the interstitial velocity of a radionuclide due to advection, K is the hydraulic conductivity, $\frac{dh}{dL}$ is the hydraulic gradient, θ is the porosity of the formation, and R is the retardation factor. Retardation factors only for the two most common isotherms namely Freundlich and Langmuir are reported herein. In Eq. (35) the retardation factor for Freundlich can be expressed as Eq. (36) [92]:

$$R = 1 + \left\{\left[\frac{BD}{\theta}\right]K_d\right\} \tag{36}$$

where R is the retardation factor, BD is the bulk density of the soil, θ is the porosity of the formation, and K_d is the partition ratio. For Langmuir, the retardation factor is according to Eq. (37) [93]:

$$R = 1 + \left(\frac{BD}{\theta}\right)\left(\frac{kb}{(1 + aCeq)2}\right) \tag{37}$$

where R is the retardation factor, BD is the bulk density of the soil, θ is the porosity of the formation, kb is the intercept, k is the maximum amount sorbed, b is the binding energy, and Ceq is the concentration at equilibrium.

11. Summary

A finite amount of radionuclides is present in virtually every environmental sphere. In general, the characteristics of radionuclides will depend on the source and their properties. A better characterization of speciation is necessary for gaining a better understanding of processes and mechanisms governing their sorption and thereby their behavior and fate in the environment. This work discussed the most significant aspects of radionuclides sorption processes and properties at the solid-water interface, the mechanisms involved, and various mathematical relationships. Furthermore, knowledge of environmental transport, environmental pathways and exposure pathways to radionuclides is an important component of any strategy to protect every biotic and abiotic environmental component. Much work remains to be done in facilitating the practical applications of the sorption–desorption characteristics of radionuclides in the environment.

Author details

Roger Saint-Fort

Address all correspondence to: rsaintfort@mtroyal.ca

Faculty of Science and Technology, Department of Environmental Science, Mount Royal University, Calgary, AB, Canada

References

[1] United States Nuclear Regulatory Commission (U.S.NRC). Protecting people and the environment. NRC: Man-made Sources. October 02, 2017

[2] Argonne National Laboratories, EVS. Human Health Fact Sheet. August 2005

[3] Sajih M, Bryan ND, Livens FR, Vaughan DJ, Descotes M, Phrommavanh V, Nos J, Morris K. Adsorption of radium and barium on geoththite and ferrihydrite: A kinetic and surface complexation modeling study. Geochemica et Cosmochimica Acta. 2014;**146**:150-163

[4] Sabine G, Louise JC, David RT, James AD, Kirk JC. Adsorption and desorption processes in subsurface reactive transport modeling. Vadose Zone Journal. 2007;**6**(3):407-435

[5] Kamil S, Patrycja B, Zofia S. Analysis of the sorption properties of different soils using vapour adsorption and potentiometric titration methods. International Agrophysics. 2016; **30**:369-374

[6] Benes P, Borovec Z, Strejc P. Interaction of radium with freshwater sediments and their mineral components:II. Kaolinite and montmorillonite. Journal of Radioanalytical and Nuclear Chemistry. 1985;**89**:339-351

[7] Ames LL, Rai D. Radionuclide interactions with soil and rock media. EPA 520/6-78-0027. Vol 1, 1978

[8] Noppadol S, Pongsakorn P. Adsorption behavior of heavy metals on various soils. Polish Journal of Environmental Studies. 2014;**3**(23):853-865

[9] Philip TU. Naturally Occurring Radioactive Materials: Principles and Practices. Advances in Environmental Science Series. Delray Beach, Florida: St Lucie Press; 1996. pp. 17-33

[10] Maiti TC, Smith MR, Laul JC. Sorption of uranium, thorium, and radium on matrices under oxic environments. Radioactive Waste Management and the Nuclear Fuel Cycle. 1989;**11**(3):269-278

[11] Laul JC, Smith MR. Disequilibrium study of natural radionuclides of uranium and thorium series in cores and briny groundwater from Palo Duro Basin, Texas. Radioactive Waste Management and the Nuclear Fuel Cycle. 1988;**11**(2):169-225

[12] Greeman DJ, Rose AW, Washington JW, Dobos RR, Giolkosz EJ. Geochemistry of radium in soils of the eastern United States. Applied Geochemistry. 1999;**14**:365-385

[13] Patterson JW, Passino R, editors. Metals speciation, separation, and recovery. In: Proceedings of the 2nd International Symposium; Rome, Italy: Lewis Publishers; 1989

[14] Von Gutten HR, Benes P. Speciation of radionuclides in the environment. Radiochimica Acta. 1995;**69**:1-29

[15] Camplin WC, Grzechnick M, Smedley CA. Methods for assessment of total dose in the radioactivity in food and the environment report. NDAWG/3/2005 Environmental Agency, Food Standadrds Agency, HPA, NII, Chilton; 2005

[16] Jones KA, Smith JG, Anderson T, Harvey MP, Brown I, Field SJ, Jones AL. Implied doses to the population of the EU arising from reported discharges from EU nuclear power stations and reprocessing sites in the years 2004 to 2008. European Commission RP 176. Available from: https://ec.europa.eu/energy/sites/ener/files/documents/176.pdf; 2013

[17] Grabianowski E. 2018. How radiation sickness works. 26 April 2011. Available from: https://science.howstuffworks.com/radiation-sickness.htm 16 January 2018

[18] Konoplev A. Fate and transport of radionuclides in soil-water environment: Review. In: Proceedings of the 19th EGU General Assembly, EGU2017; 23–28 April 2017; Vienna, Austria. p. 3231

[19] Pfingsten W, Herdamann J, Perrochet P. Radionuclide release and transport from nuclear underground tests performed at Mururoa and Fangataufa - predictions under uncertainty. Journal of Contaminant Hydrology. 2001;**47**(2–4):349-363

[20] Verkhovskaya IN, Vavilov PP, Maslov VI. The migration of natural radioactive elements under natural conditions and their distribution according to biotic and abiotic environmental components. In: Proceedings of international symposium on radioecological concentration processes. Oxford: Pergamon Press; 1967. pp. 313-316

[21] Ames LL, McGarrah JE, Walker BA. Sorption of uranium and radium by biotite, muscovite and phlogeopite. Clays and Clay Mineral. 1983;**31**:343-351

[22] Bors J, Erten H, Martens R. Sorption studies of radioiodine on soils with special references to soil microbial biomass. Radiochimica Acta. 1991;**53**:317-321

[23] Mishra S, Arac H, Zamoslyan PV, Ishikawa T, Yonchara H, Sahoo SK. Sorption-desorption characteristics of uranium, cesium and strontium in typical podzol soils from Ukraine. Radiation Protection Dosimetry. 2012;**152**:238-242

[24] Muhammad NIA, Nurul II, Khalik W, Ahmad S, Zaini H. Effectiveness of mineral soil to adsorb the natural occurring radioactive material (norm), uranium and thorium. In: AIP Conference Proceedings. Vol 1659. p. 05005, April 2015. DOI: 10.1063/1.4916875

[25] Jones MJ, Butchins LJ, Charnock JM, Patrick RAD, Small JS, Vaughan DJ, Wincott PL, Livens FR. Reactions of radium and barium with the surfaces of carbonate minerals. Applied Geochemistry. 2011;**26**:1231-1238

[26] Sprynsky M, Buszewski B, Terzyk AP, Namiešick J. Study of the selection mechanism of heavy metal (Pb^{2+}, Cu^{2+}, Ni^{2+}, and Cd^{2+}) adsorption on clinoptilotite. Journal of Colloid and Interface Science. 2006;**304**:21-28

[27] Lawson P, Steritt RM, Lester JN. Adsorption and complexation mechanisms of heavy metal uptake in activated sludge. Journal of Chemical Technology and Biotechnology. 1984;**34B**:253-262

[28] Bassot S, Mallet C, Stammose D. Experimental study and modeling of radium sorption onto goethite. Mat. Res. Soc. Symp. Proc. Vol. 663–1081. Symposium-Scientific Basis for Nuclear Waste Management XXIV; 2001

[29] Saxena S, Prasad M, D'Souza SF. Radionuclide sorption onto low-cost mineral adsorbent. Industrial and Engineering Chemistry Research. 2006;**45**:9122-9128

[30] Saint-Fort R. Sulfolane attenuation by surface and subsurface soil matrices. Journal of Environmental Science and Health Part A. 2006;**41**:1211-1231

[31] Kim YH, Yiacoumi S, Tsouris C. Surface sharges accumulation on particles containing radionuclides in open air. Journal of Environmental Radioactivity. 2015;**143**:91-99

[32] Staunton S, Haudin CS, Wang G, Shaw G. Biophysico-Chemical Processes of Heavy Metals and Metalloids in Soil Environments Chapter 13. Vol. 1. John Wiley & Sons; 2007. DOI10.1002/9780470175484

[33] Runde W. Geochemical interactions of actinides in the environment. In: Viney KM, editors. Geochemistry of Soil Radionuclides. SSSA Special Publication. 2002;**59**:21-45

[34] Walker ME, McFarlane J, Glasgow DC, Chung E, Taboada-Serrano P, Yiaccoumi S, Tsouris C. Influence of radioactivity on surface interaction forces. Journal of Colloid and Interface Science. 2010;**350**:595-598

[35] Kim Y-h, Yiacoumi S, Tsouris C. Surface charge accumulation of particles containing radionuclides in open air. Journal of Environmental Radioactivity. 2015;**143**:91-99

[36] Dyer A, Pillinger M, Newton J, Harijula R, Möller T, Amin S. Sorption behavior of radionuclides on crystalline synthetic tunnel manganese oxides. Chemistry of Materials. 2000;**12**:3798-3804

[37] Chen X, Wright JW, Concha JL, Peurring LM. Effects of pH on heavy metal sorption on mineral apatite. Environmental Science and Technology. 1997;**31**:624-631

[38] Anderson K, Torstenfelt B, Allard B. Sorption of Radionuclides in Geologic Systems. Göteborg, Sweden: Chalmers University of Technology; 1983. pp. 6-15

[39] Choppin GR, Morse JW. Laboratory studies of actinides in marine systems. Environmental Research on Actinide Elements. Office of Scientific and technical Information, DOE; 1987. pp. 49-72

[40] Wang G, Hua Y, Su X, Komarneni S, Ma S, Wang Y. Cr(VI) adsorption by montmorillonite nanocomposites. Applied Clay Science. 2016;**124–125**:111-118

[41] Schnitzer M, Kerndorff H. Reactions of fulvic acids with metal ions. Water, Air, and Soil Pollution. 1981;**15**:97-101

[42] Reuter JH, Perdue EM. Importance of heavy metal-organic matter interactions in natural waters. Geochimica et Cosmochimica Acta. 1977;**41**:325-334

[43] Noroozi B, Sorial GA. Applicable models for multi-component adsorption of dyes: A review. Journal of Environmental Sciences. 2013;**25**(3):419-429

[44] Abdehagh N, Tezel FH, Thibault J. Multicomponent adsorption modeling: Isotherms for ABE model solutions using activated carbon F-400. Adsorption. 2016;**22**:357-370

[45] Morbidell M, Servida A, Storti G, Carrã S. Simulation of multicomponent adsorption beds. Model analysis and numerical solution. Industrial and Engineering Chemistry Fundamentals. 1982;**21**:123-131

[46] Dervanoski LA, Souza SM, Luz CD, Rezende RV, Souza AU. Multicomponent adsorption and desorption of BTX compounds using coconut shell activated carbon: Experiments, mathematical modeling, and numerical simulation. Industrial and Engineering Chemistry Research. 2013;**52**:7896-7911

[47] Sulaymon AH, Ahmed KW. Competitive adsorption of fulfural and phenolic compounds onto activated carbon in fixed bed column. Environmental Science and Technology. 2008; **42**:392-397

[48] Hu X, Do DD. Multicomponent adsorption kinetics of hydrocarbons onto activated carbon: Effects of adsorption equilibrium equations. Chemical Engineering Science. 1991;**47**:1715 1725

[49] Gutierrez M, Fuentes HR. Modeling adsorption in multiple component systems using a freundlich-type isotherm. Journal of Contaminant Hydrology. 1993;**14**(3–4):247-260

[50] Travis CC, Etnier EL. A survey of sorption relationships for reactive solutes in soil. Journal of Environmental Quality. 1981;**10**(1):8-17

[51] Melikhov IV, Berdonosov SS, Znamenskaya IV, Berdonosova DG. Kinetic model for radionuclide sorption. Radiochemistry. 2007;**50**(4):338-344

[52] Xu S, Wörman A. Implications of sorption kinetics to radionuclide migration in fractured rock. Water Resources Research. 1999;**35**:3429-3440

[53] Ciffroya P, Garnierb J-M, Pham MK. Kinetics of the adsorption and desorption of radionuclides of co, Mn, Cs, Fe, Ag and cd in freshwater systems: Experimental and modelling approaches. Journal of Environmental Radioactivity. 2001;**55**:71-91

[54] Periáñez R. Testing the behaviour of different kinetic models for uptake/release of radionuclides between water and sediments when implemented in a marine dispersion model. Journal of Environmental Radioactivity. 2004;**71**:243-259

[55] Lagergren S. Zur theorie der sogenannten adsorption gelöster stoffe. Kungliga Svenska Vetenskapskademiens, Handlingar. 1898;**24**(4):1-39

[56] Roginsky SZ, Zeldovich J. Uber den mechanismus der katalytischen von CO an MnO_2. Acta Physicochim, USSR. 1934;**1**:554-594

[57] Chien SH, Clayton WR. Application of Elovich equation to the kinetics of phosphate release and sorption in soils. Soil Science Society of America Journal. 1980;**44**:265-268

[58] Enfield CG. Rate of phosphorous sorption by five Oklahoma soils. Soil Science Society of America Proceedings. 1974;**38**:404-407

[59] Leistra M, Dekkers WA. Compound effects of adsorption kinetics on pesticide movement in soils. Journal of Soil Science. 1974;**28**:340-350

[60] Hornsby AG, Davidson JM. Solution and adsorbed fluometuron concentration distribution in a water-saturated soil: Experimental and predicted evaluation. Soil Science Society of America Proceedings. 1973;**37**:823-828

[61] Smith PA, Degueldre C. Colloid-facilitated transport of radionuclides through fractured media. Journal of Contaminant Hydrology. 1993;(1–4):143-166

[62] Chang CF, Chang CY, Höll W, Ulmer M, Chen YH, Gross HJ. Adsorption kinetics of polyethylene glycol from aqueous solution onto activated carbon. Water Research. 2004; **38**(10):2559-2570

[63] Shawan T. Sorption kinetics: Obtaining a pseudo-second order rate equation based on a mass balance approach. Journal of Environmental Chemical Engineering. 2014;**2**(2):1001-1006

[64] Rizk HE, Attallah MF, Ali AMI. Investigations on sorption performance of some radionuclides, heavy metals and lanthanides using mesoporous adsorbent material. Journal of Radioanalytical and Nuclear Chemistry. 2017;**314**:2475-2487

[65] Mohan D, Gupta VK, Srivastava SK, Chandler S. Kinetics of mercury adsorption from wastewater using activated charcoal derived from fertilizer waste. Colloids and Surfaces A: Physicochemical and Engineering Aspects. 2001;**177**(2–3):169-181

[66] Valderrama C, Gamisans X, Heras X, Farran A, Cortina JL. Sorption kinetics of polycyclic aromatic hydrocarbons removal using granular activated carbon: Intraparticle diffusion coefficients. Journal of Hazardous Materials. 2008;**157**(2–3):386-396

[67] Novak LT. A method for the study of interphase mass transfer at very Reynolds number in packed bed. American Institute of Chemical Engineers Journal. 1977;**23**:398-400

[68] Vermeulen T, Heister NK. Adsorption and ion exchange. In: Perry RH, Chilton CH, editors. Chemical Engineers Handbook. 5th ed. New York: McGraw-Hill

[69] Gromiec MJ, Matsui S. Mass transfer models of sorption and desorption of radionuclides by aquatic plants. Retrieved January 22, 2018. Available from: http://hydrologie.org/redbooks/a125/iahs_125_0068.pd

[70] Holloran M. Model for instream regulation of radioisotopes and heavy metals in riverine waters subjected to a uranium mill discharge. Hydrobiologia. 1982;**91**(1):175-188

[71] Novak LT, Adriano DC. Phosphorous movement in soils: Soil-orthophosphate reactions kinetics. Journal of Environmental Quality. 1975;**4**:261-266

[72] U.S. Nuclear Regulatory Commission. Radionuclide-Chelating Agent Complexes in Low-Level Radioactive Decontamination Waste; Stability, Adsorption and Transport Potential. Pacific Northwest National Laboratory. NUREG/CR- 6758, PNNL- 13774, February 2002

[73] Weber WJ, Benjamin MV. Fundamental concepts for application of activated carbon in water and wastewater treatment. In: Suffet IH, McGuire MJ, editors. Activated Carbon Adsorption of Organics from Aqueous Phase Vol. 1. Michigan: Ann Arbor Sc. Publisher; 1978

[74] Cabal B, Ania CO, Parra JB, Pis JJ. Kinetics of naphthalene adsorption on an activated carbon: Comparison between aqueous and organic media. Chemosphere. 2009;**76**:433-438

[75] Tangestani F, Rashidi A, Mallah M-H. The kinetic study of cesium, strontium, and rubidium Radionuclide's adsorption from synthetic and natural wastes via the mag-molecular process. Water, Air, and Soil Pollution. 2017;**228**:16-29

[76] Smith JT, Comans RNJ. Modelling the diffuse transport and remobilization of ^{137}Cs in sediments. The effect of sorption kinetics and reversibility. Geochimica et Cosmochimica Acta. 1996;**60**:995-1004

[77] Ittner T, Torstenfelt B, Allard B. Diffusion of strontium, iodine, and cesium in granite rock. Radiochimica Acta. 1990;**49**:101-106

[78] Giles H, MacEwan H, Nakhawa SN, Smith D. Studies in adsorption. Part XI.* a system of classification of solution adsorption isotherms, and its use in diagnosis of adsorption mechanisms and in measurement of specific surface areas of solids. Journal of the Chemical Society. 1960:3973-3993

[79] Langmuir I. The adsorption of gases on plane surface of glass, mica and platinum. Journal of the American Chemical Society. 1926;**40**:1361-1368

[80] Freundlich H. Colloid and Capillary Chemistry. London: Matheun; 1926. pp. 110-134

[81] Manes M, Hofer LJE. Application of the Polanyi adsorption potential theory to adsorption from solution on activated carbon. The Journal of Physical Chemistry. 1969;**73**:584-590

[82] Dubinin MM. The potential theory of adsorption of gases and vapors for adsorbents with energetically non-uniform surfaces. Chemical Reviews. 1960;**60**(2):235-241

[83] Polanyi M. Neues über adsorption und ursache der Elektrochem. 1920;**26**:370-374

[84] Dubinin MM, Radushkevich LV. Doklady Akademii Nauk SSSR. 1966;**55**:331-335

[85] Brooks RH, Corey AT. Hydraulic properties of porous media. Hydrology Papers 3. Fort Collins, CO: Colorado State University. 1964. p. 37

[86] Charbeneau RJ, Daniel DE. Contaminant transport in unsaturated flow. In: Maidment DR, editor. Handbook of Hydrology, Chapter 15. New York: McGraw-Hill; 1993

[87] Short TE. Modeling of processes in the unsaturated zone. In: Loehr RC, Malina JF Jr., editors. Land Treatment: A Hazardous Waste Management Alternative, Water Resources Symposium No. 13. U. S. Environmental Protection Agency and University of Texas at Austin; 1986. pp. 211-240

[88] Bedient PB, Rifai HS, Newell CJ. Groundwater Contamination: Transport and Remediation. 2nd ed. Upper Saddle River, NJ: Prentice Hall; 1999

[89] Testoni R, Levizzari R, De Salve M. Radionuclides transport phenomena in vadose zone. International Journal of Physical, Nuclear Science and Engineering. 2014;**8**(3):31-37

[90] Radioactive Waste Management. Geological Disposal: Behaviour of Radionuclides and Non-Radiological Species in Groundwater Status Report. December 2016. NDA Report no. DSSC/456/01

[91] Wörman Ae, Geier J, Xu S. Modelling of radionuclide transport by groundwater motion in fractured bedrock for performance assessment purposes. October 2003. Swedish nuclear power. Inspectorate (SKI) Report 2004:14

[92] Martinez D, Mascioli S, Bocanegra E. Determination of Zn partition coefficient and simulation of reactive transport from landfills in Mar Del Plata, Argentina. Environmental Geology. 2006;**51**:463-469

[93] Fetter CW. Contaminant Hydrogeology. New York: Macmillan; 1993

Chapter 7

Complexation Study of Uranyl Ion with Dissolved Organic Matter in Natural Freshwater by Fluorescence Quenching Techniques

Bingqi Zhu and David Ryan

Additional information is available at the end of the chapter

http://dx.doi.org/10.5772/intechopen.72861

Abstract

The environmental problem of uranium waste has attracted global attention, and the investigation of its migration behavior in the environment has become an important topic. Uranyl ion is the most distributed form of uranium in water, and its mobility is highly affected by dissolved organic matter (DOM) due to its complexation in aquatic systems. In this study, DOMs in a variety of water samples from the Merrimack Valley of Massachusetts in the USA were studied using fluorescence excitation-emission matrix (EEM) techniques and regional integration analysis (RIA) data treatment method. RIA divided an EEM of DOM into five regions according to its fluorescence features and categorized them as humic-acid-like, fulvic-acid-like and amino-acid-like. Fluorescence quenching techniques were used to study the complexation properties of DOM with uranyl ion in aquatic systems at pH 3.5. Intense peaks in regions III and V were found to be quenched during the titration by uranyl ion. Results obtained showed that the stability constants (log K) were 4.01–4.19 and 3.83–3.97 for regions III and V, respectively. This study provides an easy, nondestructible and effective approach for studying the complexation of uranyl ion with DOM by applying the RIA method with fluorescence quenching.

Keywords: dissolved organic matter, uranyl ion, excitation emission matrix, fluorescence quenching, regional integration analysis

1. Introduction

The mining and refining activities of radionuclides for military and energy application have been dramatically increased since the past century. Uranium is one of the important radionuclides that have been extensively used, and its production capacity will increase to 122,260 tons/year by 2020, which is a 40% increase compared to 2010 [1]. Radioactive waste as a by-product during the production has resulted in environmental contamination hazards to ecosystems and human heath from both radioactivity and toxicity [2]. The majority of uranium can be found in the oxidized U (VI) state, while reduced U (IV) is also possible in water, soil and rocks. When dissolved in aquatic systems, uranyl ion (UO_2^{2+}) exists as the free ion in acidic environments, and it may form uranyl hydroxyl and uranyl carbonate ion pairs when in neutral and basic environments [3, 4]. The migration of uranyl in aquatic systems can be affected by a wide variety of factors and dissolved organic matter (DOM) plays an important role in the migration of uranyl ion [5]. Many techniques, including chromatography and mass spectrometry, have been used to study DOM and its interactions with uranium in the environment [5–8]. Fluorescence spectroscopy became an effective technique for studying DOM due to the existence of abundant aromatic groups and conjugated structures in DOM, which provide relatively good fluorescence characteristics [9, 10]. The fluorescence signal from DOM decreases when it forms complexes with certain metal ions, and based on this phenomenon, the fluorescence quenching technique was developed to study metal binding behavior of DOM [11, 12]. The advantages, such as good sensitivity, easy sample preparation and non-destruction to the sample, make the fluorescence quenching technique an effective approach to quantitatively studying the interaction between DOM and metal ions in aquatic samples.

The fluorescence excitation emission matrix (EEM) has been widely used in recent work [13]. It can provide detailed information of analytes in a three-dimensional spectrum by collecting signals from emission scans at different excitation wavelengths within a certain range. In order to interpret the EEM spectra, Chen developed the regional integration analysis (RIA) method [9]. In the RIA method, an EEM is divided into different regions according to fluorescence spectral characteristics and quantitative analysis is conducted by integrating the volume under the peaks in those regions. This RIA method has been used in the study of DOM previously [14, 15].

In this study, a new approach was developed by applying the RIA method to fluorescence quenching techniques to investigate the complexation behavior between uranyl ion and DOM in natural freshwater samples. This procedure was employed to aid in elucidating the binding mechanism of DOM with pollutants such as radioactive metal ions in natural aquatic systems, which will further improve understanding of migration of metal ions in the environment. Therefore, this study was aimed at investigating the fluorescence properties of DOM in natural freshwater samples and the complexation properties between uranyl ion and DOM using the fluorescence quenching technique coupled with the RIA data treatment method.

2. Experimental section

2.1. Apparatus

Fluorescence data were acquired by a Perkin Elmer LS55 spectrofluorometer. Instrumental parameters were set as follows: excitation slit—7 nm; emission slits—7 nm; excitation wavelength—200–400 nm (step—5 nm); emission wavelength—250–550 nm (step—0.5 nm); scan speed—600 nm/min; solution temperature—25°C. A WTW pH meter was used to monitor the pH.

2.2. Samples and reagent

Natural water samples were collected from the Merrimack River valley in Massachusetts, USA. The five sampling locations were Haggetts Pond (Andover, MA), Concord River (Lowell, MA), Mascuppic Lake (Tyngsboro, MA), Merrimack River Lowell site (Lowell, MA) and Merrimack River Estuary site (Newburyport, MA). Five liters acid-washed polyethylene bottles were used to store and transfer water samples. Before laboratory analysis, samples were filtered through 0.22 μm Whatman (Maidstone, UK) nylon membranes and stored in Corning polystyrene roller bottles in a refrigerator at 4°C. Uranium atomic absorption standard solution (1000 μg/mL) was purchased from Ricca Chemical (Arlington, TX). The deionized water was obtained from an Elga Purelab water purification system.

2.3. Fluorescence titration experiments

In the titration experiments, a series of accurately measured amounts of uranyl ion standard solution were titrated into a beaker containing a known volume of natural water sample. EEMs of the water sample were collected before the titration and after each addition of uranyl ion solution. The concentration range of uranyl ion was from 0 to 0.68 mmol/L by the end of the titration. Measurements were conducted three times at 25°C (±1°C) and at pH 3.50 (±0.01) and the same procedure was applied to each of the natural water samples. The experimental pH was maintained at 3.50, because most of the uranyl exists as free uranyl ion at this pH value [16].

3. Theory and analysis

3.1. Theories of RIA

Absorption of light energy may lead to transitions of electronic, vibrational and rotational states of a molecule. As illustrated in **Figure 1**, a simplified Jablonski diagram, an electron may be excited from the ground singlet state (S_0) to different vibrational and rotational energy levels of the first excited singlet state (S_1) by absorption of light at different wavelengths. Excited electrons in S_1 may lose energy via vibrational relaxation and return to the lowest vibrational energy state in S_1. When electrons return to S_0 from S_1, they may go back to different vibrational or rotational states by emitting light at different wavelengths within a certain range. Therefore, a fluorophore can be excited by light over a certain range of wavelengths and radiate in another wavelength range. A fluorescence peak follows a near Gaussian distribution [18]. The volume of the peak for RIA in an EEM can be calculated by the integrated area of a fluorophore peak over a range of excitation wavelengths (y axis of EEM) and a range of emission wavelengths (x axis of the EEM) times the fluorescence intensity plotted on the z axis. Based on this theory, quantitative analysis was achieved by the RIA method.

Figure 2 shows a typical EEM of Concord River water sample. Two distinct peaks were observed: one in region III (peak center at excitation—245 nm; emission—440 nm) and one in region V (peak center at excitation—330 nm; emission—440 nm), which were recognized as the fulvic-acid-like region and humic-acid-like region. Amino-acid-like and microbial by-product

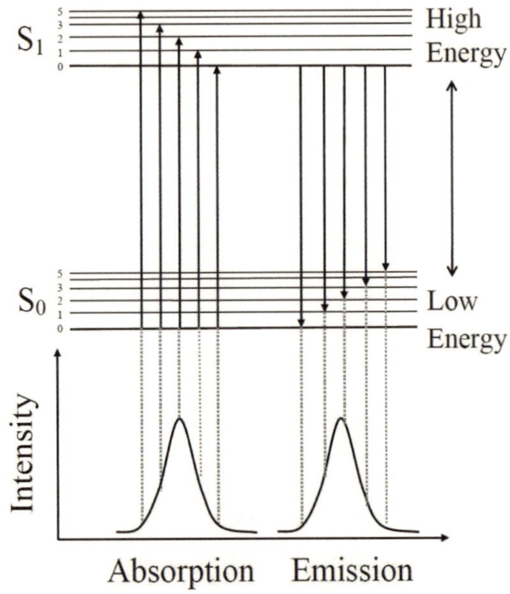

Figure 1. Jablonski diagram for fluorescence [17].

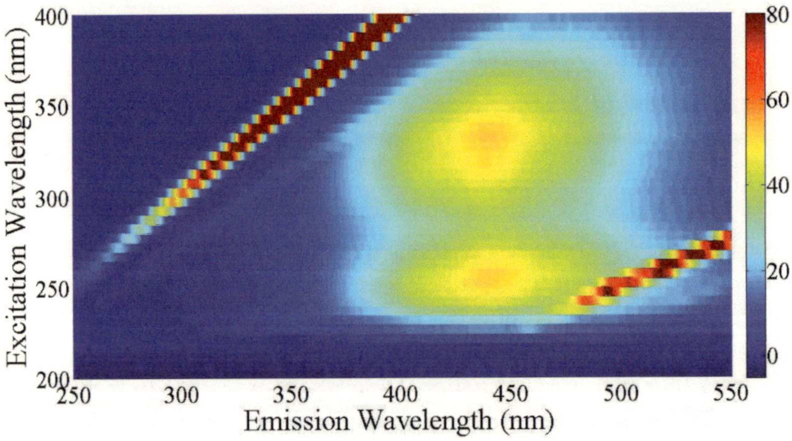

Figure 2. Fluorescence EEM spectrum of Concord River water sample.

related substances are categorized in region IV and were absent. This region includes tryptophan (excitation—280 nm; emission—348 nm), tyrosine (excitation—274 nm; emission—303 nm), and phenylalanine (excitation—257 nm; emission—282 nm).

3.2. RIA data treatment method

According to the RIA method [9], five regions were defined, as demonstrated in **Table 1**. The peak volume was calculated by integration of the area multiplied by the height of the area. A blank EEM spectrum of deionized water was subtracted from sample EEMs to remove any Raman scattering effect. The first-order Rayleigh scattering effect was eliminated by setting

Region	Excitation wavelength (nm)	Emission wavelength (nm)	Description
I	200–250	280–330	Amino-acid-like
II	200–250	330–380	Amino-acid-like
III	200–285	380–550	Fulvic-acid-like
IV	250–400	280–380	Amino-acid-like and soluble microbial by-product-like
V	285–400	380–550	Humic acid-like

Table 1. RIA definition of five regions in an EEM [9, 17].

values in that region to zeros. The second-order Rayleigh scattering effect was eliminated by an interpolation method, which allowed the replacement of the old data by new values in the second-order Rayleigh scattering region until a smooth surface was obtained. MATLAB software (version 2010b, Mathworks, Natick, MA) was used to conduct the RIA.

3.3. Curve fitting modeling

The curve fitting was carried out based on the nonlinear regression analysis of the experimental titration curves and the complexation model [12]. The modeling has been described in our previous study [17]. Briefly, based on the assumption of 1:1 complexation ratio between metal and ligand, a reaction can be described as follows:

$$M + L \leftrightarrow ML \quad (1)$$

where M is the free metal ion; L is the metal-free ligand site of a fluorescent species; and ML is the metal-bound species.

A stability constant for the reaction is represented in Eq. (2), and mass balances for metal and ligand can be represented in Eqs. (3) and (4), respectively. Fluorescence intensity is proportional to the concentration of fluorophores during the titration, and this phenomenon can be represented by the Ryan-Weber model [11, 12], Eq. (5). By combining and rearranging the above equations, florescence intensity I and C_M are represented in Eq. (6):

$$K = [ML]/[M][L] \quad (2)$$

$$C_M = [M] + [ML] \quad (3)$$

$$C_L = [L] + [ML] \quad (4)$$

$$[ML]/C_L = (I_L - I)/(I_L - I_{res}) \quad (5)$$

$$I = [(I_{res} - 100)/2K\,C_L]\{(K\,C_L + K\,C_M + 1) - [(K\,C_L + K\,C_M + 1)^2 - 4K2\,C_L\,C_M]^{1/2}\} + 100 \quad (6)$$

where K is the conditional stability constant; $[ML]$ is the concentration of metal-bound species; $[M]$ is the concentration of free metal ion; $[L]$ is the concentration of metal-free ligand; C_M is the total concentration of metal in the solution; C_L is the total concentration of ligand in the solution; I_L is the fluorescence intensity of ligand without any addition of metal; I is the fluorescence intensity of unbound ligand during the titration process; and I_{res} is the residual fluorescence intensity at the end of the titration.

With the fluorescence quenching data obtained in the experiment, nonlinear regression of the titration curve was conducted using Eq. (6) and K, C_L and I_{res} values were obtained.

4. Results and discussion

4.1. Fluorescence quenching analysis

The fluorescence quenching phenomenon was observed when aliquots of uranyl ion standard solution were titrated into natural water samples. As shown in **Figure 3**, the two distinct peaks in region III and region V were found to be quenched.

As the incremental concentration of uranyl ion in solution increased, the fluorescence signal of each region was decreased to varying degrees. The volumes of the solution were calculated after each addition of uranyl ion, and based on the change of solution volumes and the amount of uranyl ion added, quenching curves were plotted. Uranyl ion binding capacities for natural water samples were studied via fluorescence quenching phenomena for these regions. **Figure 4** shows the fluorescence quenching curves of the Concord River water sample titrated with uranyl ion and good reproducibility of the quenching curves were obtained in regions III and V. Fluorescence intensities during the titration were adjusted to a percentage relative to the initial intensity of the sample before the titration started. The fluorophores in region V show a higher residual fluorescence value than in region III, which might be caused by several factors: (1) the binding site associated with fluorophores in region V has a lower binding affinity for uranyl ion, which makes it less quenchable. (2) The fluorophores in region V may have a high fluorescence efficiency in their bound form, which enables the complex to absorb and emit significantly even when it is bound with metal. (3) There could be nonbinding fluorescent DOM in this spectral region producing a high residual fluorescence. Compared with the Concord River water, similar trends were observed for other natural water samples. This related trend represented a relatively similar DOM composition between these different samples.

Nonlinear regression analysis using the Ryan-Weber equation [11, 12] was conducted, and the results are presented in **Table 2**. In another experiment, the interaction between Cu^{2+} and tryptophan was studied, and a very good agreement was obtained between our results and literature reports by using the RIA method coupled with the fluorescence quenching technique. This proved the reliability of the application of the RIA method to fluorescence quenching analysis.

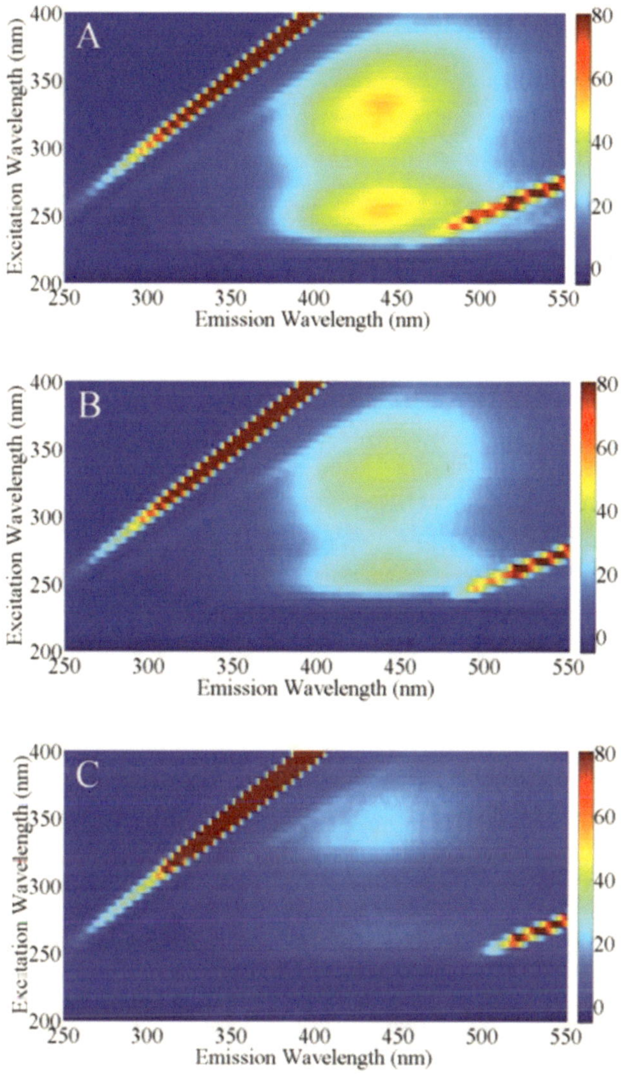

Figure 3. Selected examples of fluorescence EEMs of the Concord River sample during the titration with uranyl ion at concentrations of: (A) 4.2 μmol/L; (B) 41.5 μmol/L; and (C) 375 μmol/L.

4.2. Effect of pH

Uranyl chemistry is highly affected by the pH environment. As shown in **Figure 5**, uranyl mainly exists as free uranyl ion in solutions that have a pH lower than 3.5. In a neutral environment, the majority of uranyl ion is in the form of a uranyl hydroxyl complex and it transforms into a uranyl carbonate complex in basic conditions [16]. Several factors should be considered in studying the complexation of uranyl ion and DOM. First, in an acidic system that has pH values lower than 3.5, the high concentration of hydrogen ion will compete with uranyl ion when

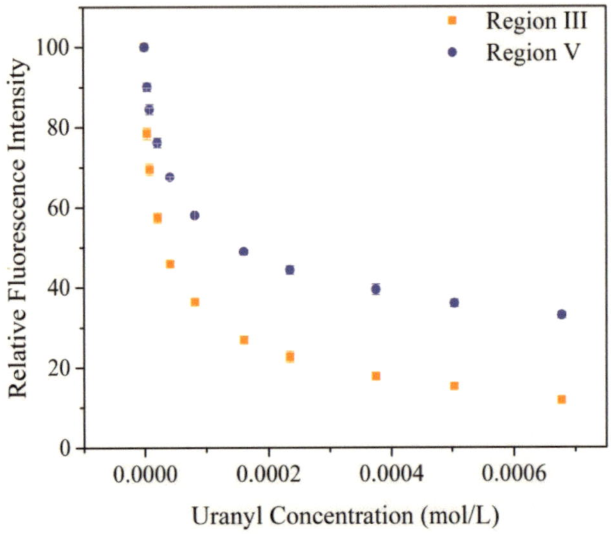

Figure 4. Fluorescence quenching curves for the Concord River water sample titrated with uranyl ion at pH 3.50.

binding with DOM. Study results showed that DOM was less quenched when the pH was lower and more when pH increased. Second, when the pH is higher than 3.5, uranyl mainly exists as uranyl hydroxyl and uranyl carbonate complexes and complexation behavior with DOM may be changed in these situations. Moreover, uranyl was found to exhibit strong fluorescence when the pH was at 5.5; this may cause difficulties in applying the fluorescence quenching techniques [3, 17]. When the pH was maintained at 3.5, uranyl existed as free uranyl ion and had a negligible fluorescence signal; therefore, this condition was selected.

Sample	Region	log *K* (±Std. Dev.)	C_L (μmol/L) (±Std. Dev.)	I_{res} (±Std. Dev.)
Haggetts Pond	Region III	4.38 (±0.03)	*	10.1 (±0.4)
	Region V	3.83 (±0.08)	*	35.5 (±4.9)
Mascuppic Lake	Region III	4.01 (±0.03)	*	3.4 (±3.34)
	Region V	3.93 (±0.07)	4.32 (±0.93)	34.2 (±1.3)
Concord River	Region III	4.26 (±0.02)	*	6.7 (±0.4)
	Region V	3.93 (±0.04)	*	23.6 (±1.7)
Merrimack River Lowell	Region III	4.49 (±0.05)	*	8.1 (±0.2)
	Region V	3.97 (±0.03)	*	28.0 (±1.7)
Merrimack River Newburyport	Region III	4.08 (±0.03)	*	3.4 (±2.3)
	Region V	3.91 (±0.04)	4.83 (±3.26)	29.3 (±0.5)

*Are the values not provided due to extremely small values obtained.

Table 2. Conditional stability constant (log *K*), residual fluorescence intensity (I_{res}) and ligand concentration (C_L) determined by the fluorescence quenching techniques and the RIA method.

http://dx.doi.org/10.5772/intechopen.72861

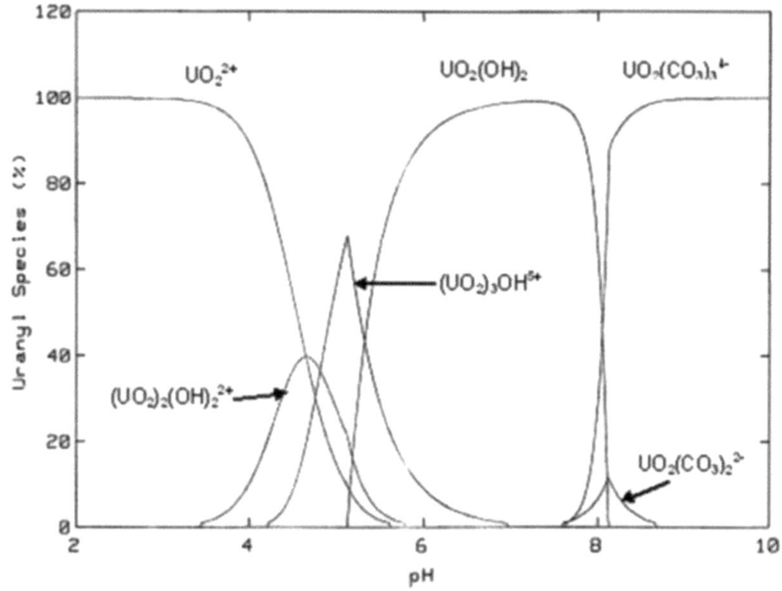

Figure 5. Percentage of chemical species for 0.8 mmol/L uranyl ion in a 0.01 mol/L ionic strength solution and 25°C in equilibrium with the atmosphere. Distributions were calculated with MINEQL + ionic equilibrium program [16].

4.3. Further study

While the fluorescence quenching technique provided an approach to understand the complexation between uranyl ion and DOM, a more complicated situation should be considered in the study of the transport or deposition of uranium with DOM in natural aquatic systems. In this study, the modeling is based on a simplified assumption of 1:1 complexation between uranyl ion and fluorescent DOM ligand. Since DOM has a complicated structure, a fluorophore may be associated with more than one site that is able to bind uranyl ion. On the other hand, other ligands that do not fluoresce may bind with uranyl and compete with fluorescent ligands. The metal ions in natural aquatic systems may compete with uranyl ion in complexation with DOM, which reduces the availability of the DOM that binds with uranyl ion. Other factors that affect the complexation of the DOM and uranyl ion in natural aquatic systems, such as pH, hardness and alkalinity, are highly dependent on the local geology, weather condition and human activity. To better understand these factors and the transport behavior of uranyl ion, further studies should be carried out.

5. Conclusion

The migration of uranium was studied by investigating the binding properties of uranyl ion and DOM. The fluorescence quenching technique coupled with the RIA data treatment was applied in this study. Different water types were studied, including lake, pond, river and estuary. Although challenges existed due to the heterogeneous and complex properties of

DOM and low C_L values obtained, this method provided reliable results in measuring stability constants. From the results, two distinct peaks were observed in region III and region V and the fluorescence quenching of these two peaks was studied. The Concord River water showed the highest stability constants followed by Mascuppic Lake, Haggetts Pond, Merrimack River Lowell and Merrimack River Newburyport. Region V was less quenchable and lower in binding ability than region III. The fluorescence quenching method is easy and nondestructive to the sample and can be used to monitor natural water quality and conduct binding studies. Further study of a more complicated modeling such as multisite binding modeling is needed to better understand the complexation of uranium with DOM in natural aquatic systems.

Author details

Bingqi Zhu[1*] and David Ryan[2]

*Address all correspondence to: bingqi_zhu@hotmail.com

1 Zhejiang Institute for Food and Drug Control, Hangzhou, Zhejiang, China

2 Department of Chemistry, University of Massachusetts Lowell, Lowell, MA, USA

References

[1] Du X, Boonchayaanant B, Wu W, Fendorf S, Bargar J, Criddle CS. Reduction of uranium(VI) by soluble iron(II) conforms with thermodynamic predictions. Environmental Science & Technology. 2011;**45**:4718-4725

[2] Taylor DM, Taylor SK. Environmental uranium and human health. Reviews on Environmental Health. 2011;**12**:147-158

[3] Vandenhove H, Hurtgen C, Payne T, Radionuclides in the environment: Uranium, 2010, Radionuclides in the Environment, West Sussex, United Kingdom: John Wiley & Sons, Ltd.

[4] Meinrath G. Uranium(VI) speciation by spectroscopy. Journal of Radioanalytical and Nuclear Chemistry. 1997;**224**:119-126

[5] Trenfield MA, McDonald S, Kovacs K, Lesher EK, Pringle JM, Markich SJ, Ng JC, Noller B, Brown PL, van Dam RA. Dissolved organic carbon reduces uranium bioavailability and toxicity. 1. Characterization of an aquatic fulvic acid and its complexation with uranium [VI]. Environmental Science & Technology. 2011;**45**:3075-3081

[6] Zhao J, Fasfous II, Murimboh JD, Yapici T, Chakraborty P, Boca S, Chakrabarti CL. Kinetic study of uranium speciation in model solutions and in natural waters using competitive ligand exchange method. Talanta. 2009;**77**:1015-1020

[7] Möser C, Kautenburger R, Philipp Beck H. Complexation of europium and uranium by humic acids analyzed by capillary electrophoresis-inductively coupled plasma mass spectrometry. Electrophoresis. 2012;**33**:1482-1487

[8] Mibus J, Sachs S, Pfingsten W, Nebelung C, Bernhard G. Migration of uranium(IV)/(VI) in the presence of humic acids in quartz sand: A laboratory column study. Journal of Contaminant Hydrology. 2007;**89**:199-217

[9] Chen W, Westerhoff P, Leenheer JA, Booksh K. Fluorescence excitation–emission matrix regional integration to quantify spectra for dissolved organic matter. Environmental Science & Technology. 2003;**37**:5701-5710

[10] Hudson N, Baker A, Reynolds D. Fluorescence analysis of dissolved organic matter in natural, waste and polluted waters—A review. River Research and Applications. 2007;**23**: 631-649

[11] Ryan DK, Weber JH. Copper(II) complexing capacities of natural waters by fluorescence quenching. Environmental Science & Technology. 1982;**16**:866-872

[12] Ryan DK, Weber JH. Fluorescence quenching titration for determination of complexing capacities and stability constants of fulvic acid. Analytical Chemistry. 1982;**54**:986-990

[13] Henderson RK, Baker A, Murphy KR, Hambly A, Stuetz RM, Khan SJ. Fluorescence as a potential monitoring tool for recycled water systems: A review. Water Research. 2009;**43**: 863-881

[14] Chai X, Liu G, Zhao X, Hao Y, Zhao Y. Fluorescence excitation–emission matrix combined with regional integration analysis to characterize the composition and transformation of humic and fulvic acids from landfill at different stabilization stages. Waste Management. 2012;**32**:438-447

[15] Yu G, Wu M, Luo Y, Yang X, Ran W, Shen Q. Fluorescence excitation–emission spectroscopy with regional integration analysis for assessment of compost maturity. Waste Management. 2011;**31**:1729-1736

[16] Zhu B, Pennell SA, Ryan DK. Characterizing the interaction between uranyl ion and soil fulvic acid using parallel factor analysis and a two-site fluorescence quenching model. Microchemical Journal. 2014;**115**:51-57

[17] Zhu B, Ryan DK. Characterizing the interaction between uranyl ion and fulvic acid using regional integration analysis (RIA) and fluorescence quenching. Journal of Environmental Radioactivity. 2016;**153**:97-103

[18] Dubnick A, Barker J, Sharp M, Wadham J, Lis G, Telling J, Fitzsimons S, Jackson M. Characterization of dissolved organic matter (DOM) from glacial environments using total fluorescence spectroscopy and parallel factor analysis. Annals of Glaciology. 2010; **51**:111-122